A MANUAL OF RADIOBIOLOGY

BIOLOGY SERIES

General Editor: R. Phillips Dales
Professor of Zoology in the University of London at Bedford College

U.S. Editor: Arthur W. Martin,
Professor of Physiology, Department of Zoology, University of Washington

---

COMPARATIVE ANIMAL CYTOLOGY AND HISTOLOGY
by Ulrich Welsch and Volker Storch

WASPS *by* J. Philip Spradbery

RESEARCH METHODS IN MARINE BIOLOGY *edited by* Carl Schlieper

DESCRIPTION AND CLASSIFICATION OF VEGETATION *by* David W. Shimwell

MARINE BIOLOGY *by* Hermann Friedrich

PRACTICAL INVERTEBRATE ZOOLOGY
by F. E. Cox, R. P. Dales, J. Green, J. E. Morton, D. Nichols, D. Wakelin

ANIMAL MECHANICS *by* R. McNeill Alexander

THE BIOLOGY OF ESTUARINE ANIMALS *by* J. Green

STRUCTURE AND HABIT IN VERTEBRATE EVOLUTION *by* G. S. Carter

MOLECULAR BIOLOGY AND THE ORIGIN OF SPECIES *by* Clyde Manwell
and C. M. Ann Baker

ANIMAL ECOLOGY *by* Charles Elton, FRS

A GENERAL ZOOLOGY OF THE INVERTEBRATES *by* G. S. Carter

ZOOGEOGRAPHY OF THE SEA *by* S. Ekman

THE NATURE OF ANIMAL COLOURS
by H. Munro Fox, FRS and H. Gwynne Vevers

THE FEATHERS AND PLUMAGE OF BIRDS *by* A. A. Voitkevich

A MANUAL
OF
RADIOBIOLOGY

by

JOHN C. STEWART

and

DAVID M. HAWCROFT

Department of Biological Studies
Lanchester Polytechnic
Coventry

UNIVERSITY OF WASHINGTON PRESS
SEATTLE

First published in Great Britain 1977

*Published in the United States of America
by the University of Washington Press in 1977*

Copyright © 1977 J.C. Stewart and D. M. Hawcroft

Library of Congress Catalog Card Number 76-56671

ISBN 0-295-95553-8

Printed in Great Britain

ACKNOWLEDGEMENTS

We are indebted to Paul Mundy and Peter Freeman of the Polytechnic Faculty of Art and Design for producing a number of diagrams. We would also like to thank Dr A. Syson and Dr A. Ecumenides for details of the experiments involving ^{22}Na and Dr R. B. Bentley (the Polytechnic Radiological Safety Officer) for his general advice.

J. C. STEWART
D. M. HAWCROFT

Lanchester Polytechnic
April, 1976

CONTENTS

Section B

EXPERIMENTS

INTRODUCTION

It is hoped that this book will have two uses. First, to provide a theoretical introduction, at intermediate level, to the various aspects of radiobiology and secondly, to provide some experiments to illustrate the use of isotopes in biology.

The introductory chapters, *Section A*, have been written to provide a theoretical grounding for degree students from which they may progress to more advanced texts. It is considered, however, that the information presented here is adequate for a very large number of students who are on courses of a non-specialist nature and in which radiobiology is a relatively small component. Students studying for G.C.E. Advanced level should also find these chapters useful.

Particular attention should be paid to Chapter 6 on safety precautions.

The experiments described in *Section B* have been devised to illustrate the different uses of radioisotopes in the various branches of biology. The experiments are written in a form for immediate use in laboratory classes but they could readily be included as practical examples of the subject in lecture courses. Some experiments are extremely simple and straightforward and could be performed by school sixth-form students using the minimum of equipment. Others are more sophisticated and require equipment normally found only in colleges or universities and a level of theoretical knowledge and practical ability only found in students at these institutions. Most of the experiments could readily be extended or made the basis of projects.

Many of the experiments have been deliberately designed to use plant material and there are two reasons for this. Plant material is easily grown in relatively large quantities with the minimum of equipment and facilities and is more easily manipulated in practical classes.

Additionally, many experiments that might be carried out with animals require the operator to have a Home Office licence and this precludes the use of live animals on a class basis.

Owing to the fact that radioisotopes are expensive, the experiments selected use a small number of labelled substrates yet cover a wide range of biological areas and radiobiological techniques.

Most of the experiments require count rate determinations in order to obtain quantitative results, and these can be carried out using Geiger-Muller systems although the use of a scintillation counter will probably improve the quality of the results. Many of the experiments could be performed using autoradiography if counting facilities are not available and this is indicated in the experimental procedure where applicable.

Each experimental description follows the same layout. The level of student knowledge is indicated to help teachers and lecturers choose experiments most suitable for their students. Similarly the length of time and facilities required are indicated. At the end of the experimental description is a complete list of requirements, together with details of plants, micro-organisms, and other items which must be prepared in advance of the experiment.

Section A

CHAPTER 1

THE NATURE OF RADIOISOTOPES

In order to understand the nature of radiations and radioisotopes it is necessary to study briefly the structure of the atom.

For our purpose it is sufficient to consider an atom as a central nucleus around which circulate one or more electrons. Atomic nuclei (except hydrogen) consist of a mixture of protons and neutrons and the atoms are characterised by the number of protons they possess. This number is termed the atomic number and it is a feature of a particular element and hence is indicative of a particular position in the periodic table.

Natural hydrogen is a mixture of three types of atom, protium (or ordinary hydrogen), deuterium and tritium. Protium (which constitutes by far the greatest proportion of terrestrial hydrogen) has a nucleus consisting of a single proton. However, one atom in six thousand has a nucleus containing a neutron as well as a proton. This form of hydrogen is termed deuterium and was separated from ordinary hydrogen in 1931. The third type, tritium has a nucleus containing one proton and two neutrons and, in contrast to deuterium, is unstable, its atom changing to helium of the same mass. During the course of this change radiation is emitted and for this reason tritium is said to be radioactive. The structures, of these three forms of hydrogen are shown in Figure 1.

These atoms are designated ^1_1H, ^2_1H and ^3_1H respectively.

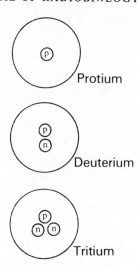

Figure 1. Hydrogen isotopes.

The superscript is the mass number and will vary because of the different number of particles in the atom and the consequent variation in its mass. The subscript is the atomic number and is a constant because all these atoms possess a single proton and because of this, occupy the same position in the periodic table. The subscript is in fact often omitted since the elemental symbol partly duplicates its role.

Although it is usual to refer to the above as isotopes of the element hydrogen, and phosphorus 31 and phosphorus 32 as isotopes of the element phosphorus, an individual atomic species should be called a nuclide. The hundred or so elements have furnished about 1,400 nuclides of which about 85% are radioactive. Generally speaking, radioactive nuclides are those with odd numbers of protons or neutrons.

Production of Radioisotopes

A number of radioisotopes occur naturally and they are either produced continuously due to bombardment of the atmosphere by cosmic rays, (e.g. ^{14}C), or they are relics from the formation of this planet (e.g. ^{40}K).

Most of the light radioisotopes used for tracer studies are made by bombardment of suitable materials with neutrons in an atomic pile. In such a reactor there is a high density of slow and fast neutrons and also abundant γ-rays. The various radioisotopes are produced principally as a result of three types of reaction.

Bombardment with a neutron and elimination of γ-radiation; an (n, γ)-reaction.

$$^{23}_{11}Na + ^{1}_{0}n \rightarrow ^{24}_{11}Na + \gamma$$

Bombardment with a neutron and elimination of a proton; an (n,p)-reaction.

$$^{32}_{16}S + ^{1}_{0}n \rightarrow ^{32}_{15}p + ^{1}_{1}p$$

Bombardment with a neutron and elimination of a deuteron or α-particle; an (n, d) reaction.

$$^{27}_{13}Al + ^{1}_{0}n \rightarrow ^{24}_{11}Na + ^{4}_{2}He$$

Unfortunately (n,γ)-reactions produce radioisotopes which are of the same element as the starting material and thus cannot be separated easily. The starting material acts as a diluent or *carrier*. However, (n,p)-reactions can give rise to products which are radioisotopes of a different element and which may be easily separated from the starting material by chemical means. These can, therefore, be produced in a very pure form.

Radioactive Decay

Radioisotopes may decay in a number of distinct ways most usually producing electrons, helium nuclei or γ-radiation. Many radioisotopes produced by neutron bombardment are β-emitters, that is, on radioactive decay they emit a β-particle (a negative electron) from the nucleus. The β-particle is produced by the decay of a neutron into both a proton and an electron, and because this additional proton remains within the nucleus the product has the same mass but a greater atomic number; the mass of the emitted β-particle being almost negligible. The product will therefore occupy a different place in the periodic table because it is of course a different element. So, for carbon 14 decay:

$$^{14}_{6}C \rightarrow {^{14}_{7}N} + \beta^-$$

The β-particle is eliminated at high speed and possesses kinetic energy; the maximum energy it possesses differing with the isotope producing it. This energy may be calculated from the mass lost in the disintegration producing it and for the above decay:

Mass for $^{14}_{6}C = 14 \cdot 003242$ atomic mass units (AMU)

Mass for $^{14}_{7}N = 14 \cdot 003074$ AMU

Difference $= 0 \cdot 000168$ AMU

From Einstein's theories of the relationship between mass and energy, one AMU is equivalent to 931 MeV of energy.

$$\beta\text{-particle energy} = 0 \cdot 000168 \times 931$$

$$= 0 \cdot 156 \text{ MeV}$$

For tritium and phosphorus 32 the values are $0 \cdot 019$ MeV and $1 \cdot 710$ MeV, respectively.

It is of some importance when measuring radiation or erecting shielding against it, that the β-particles produced by ^3H and ^{14}C are weak with little penetrating power ('soft' β-particles), but those produced by ^{32}P are more energetic ('hard' β-particles) and have greater penetrating power.

It is also possible for radioactive isotopes to decay in ways yielding other types of radiation, either instead of, or in addition to, the β-particle emission. Since these are of lesser importance in biology the physical basis of their origin is not discussed. Some properties of these radiations, however, are covered in the next chapter.

It is of fundamental importance to the use of radioisotopes in scientific investigations, that the rate of radioactive decay is independent of physical or chemical conditions. It is a characteristic of the radioisotope of the element in which the change occurs and so the decay of, say ^{32}P in PO_4^{3-}, ATP and DNA will be constant. The rate of decay is conveniently expressed numerically as the time in which half the mass of any given quantity of the element disappears, that is, the *half-life* of the isotope. The size of the given quantity does not matter, one kilogram will decay to one half kilogram in the same time as one gram decays to one half gram. The decay of a sample of an isotope is therefore exponential and hence after two half-lives, 25% of the original activity remains, after three half-lives, 12·5% remains. A semi-logarithmic graph illustrating this for ^{32}P is given in the appendices.

Obviously, all the radioactive atoms of a given nuclide do not decay at the same time; individual decays occur entirely at random and so can be defined by statistical laws. The decay constant (λ) gives a numerical indication of the rate of decay of the radioactivity in a sample and the value of λ is obviously lower for isotopes with long half-lives and higher for ones with short half-lives. The activity of a specimen is the number of disintegrations in a unit of time.

It is worth bearing in mind at this point that since radioactive isotopes are detected by the radiation they emit, the total number of radioactive atoms in a sample of an isotope of long half-life is greater than the number of radioactive atoms in a sample of the same activity containing a radioisotope of short half-life. Thus one radioactive unit (one Curie) of ^{14}C (half-life 5,570 years) weighs 0·22 g whereas one Curie of ^{131}I (8·06 days) weighs 8·08 μg.

There are of course a number of equations defining and relating these parameters. For instance, the number of atoms N, left after time t, is given by the equation:

$$N = N_0 e^{-\lambda t}$$

Where N_0 is the original number of nuclei. Similarly, for the activity A, of a specimen after time t

$$A = A_0 e^{-\lambda t}$$

where A_0 is the original activity and e the base of natural logarithms (2·7183).

The half-life ($t_{\frac{1}{2}}$), the time taken to halve the number of atoms of the radioisotope or the activity of it, is the time when

$$\frac{N}{N_0} = \frac{A}{A_0} = \frac{1}{2}$$

and substituting in either of the above equations it can be shown that

$$t_{\frac{1}{2}} = \frac{\log_e 2}{\lambda} = \frac{0 \cdot 693}{\lambda}$$

The half-lives and decay constants of the radioisotopes most commonly used in biological investigations are:

	$t_{\frac{1}{2}}$	λ
$^{3}_{1}H$	12·5 years	$5 \cdot 63 \times 10^{-2}$ yr^{-1}
$^{14}_{6}C$	5,570 years	$1 \cdot 24 \times 10^{-4}$ yr^{-1}
$^{32}_{15}P$	14·3 days	$17 \cdot 67$ yr^{-1}

Thus tritium and carbon 14 have long half-lives and can be stored for long periods of time. The short half-life of ^{32}P presents problems discussed later.

Specific Activity

If a radioisotope is not carrier-free it will be mixed homogeneously with a certain amount of the stable isotope of the same element. In order to show the relative abundance of the radioisotopes and the stable isotopes, the term *specific activity* is used.

The specific activity is the amount of radioactivity in a given weight of material and is usually expressed as Curies, disintegration rate or count rate, per unit mass of element.

The Standard of Radioactivity

The standard used for measuring radioactivity is the Curie (Ci), and this was originally defined as the activity of one gram of pure radium-226. However, to calculate this activity the half-life must be accurately known and since this is likely to change as the techniques for measuring it become more refined, the Curie was arbitrarily fixed as $3 \cdot 700 \times 10^{10}$ disintegrations per second (dps) by the Joint Commission of the International Union of Pure and Applied Chemistry and the Union of Pure and Applied Physics in 1950. For tracer studies the microcurie (μCi; 10^{-6} Curies) is often used.

The Curie may be replaced in the future by the Becquerel (Bq), when $1Bq = 2 \cdot 703 \times 10^{-11}$ Ci i.e. 1 disintegration per second.

Problems With Short Half-life Isotopes

As a final point in this chapter it should be emphasised that the experimental use of isotopes with short half-lives can introduce two main complications. It is often necessary to calculate the amount of activity left in a sample which has been purchased some time earlier, in order to determine the correct quantity to use in the experiment. This can be done using the formula given earlier.

$$A = A_o e^{-\lambda t}$$

For example, if one millicurie of ^{32}P was purchased, after $14 \cdot 3$ days the amount of activity remaining can be calculated using the decay constant $4 \cdot 84 \times 10^{-2}$ day^{-1}.

$$A = 1 \times 2 \cdot 7183^{-(4 \cdot 84 \times 10^{-2})(14 \cdot 3)}$$
$$= 2 \cdot 7183^{-0 \cdot 69}$$
$$= 0 \cdot 500 \text{mCi}$$

Alternatively and more conveniently, one can use a decay graph similar to that for ^{32}P included as Appendix 6.

An additional problem, particularly for long-term experiments, or experiments in which great accuracy is required, is a need to correct the results for the decay occurring during the course of the experiment or even while counting different experimental samples. None of the experiments in this book require such corrections.

CHAPTER 2

RADIATION CHARACTERISTICS

Spontaneous disintegrations of atomic nuclei give rise to radiations of different types. From our point of view the most important of these are α-particles, β-particles and γ-radiations and we can usefully consider some properties of these before discussing their importance in biological experiments.

Biologically Useful Types of Radioactive Isotopes

Isotopes which emit α-particles are not normally used as tracers in biological investigations because they belong to elements of high atomic number, that is, above 82. Such elements are rarely important metabolically, though some studies of the uptake of heavy elements by plants have been made.

The most important isotopes in biological investigations are those which emit β-particles because three of the isotopes most commonly used in biological experiments, ^3H, ^{14}C and ^{32}P, are in this group.

A number of isotopes with some biological significance emit γ-radiation and perhaps the most often used are ^{59}Fe, ^{24}Na, ^{125}I and ^{131}I.

Nature of the Emissions

The α-particle is relatively massive as it consists of a helium

10

nucleus, and it also has a double positive charge. The particles move relatively slowly, but due to their mass they have high momentum and will travel in straight lines and not be easily deflected from their path. Normally they are only deflected by direct collision with a nucleus.

β-particles are electrons of nuclear origin and they may be either positively (e.g. ^{22}Na) or negatively (e.g. ^{14}C) charged, although the positively charged ones (positrons) have only a transient existence. β-particles are emitted from the nucleus at high speed, normally approaching that of light. They have little mass (1/7400 the mass of the α-particle) and because of this they are relatively easily deflected from their path and as a consequence they do not travel in straight lines.

γ-radiation is sometimes emitted during the course of radioactive decay if one of the intermediate stages in the decay sequence has too much energy to maintain stability; the excess energy may then be emitted as γ-radiation. γ-radiation is a true radiation of short wavelength and is, therefore, found as a component of the electromagnetic spectrum. It has no mass and travels in straight lines at the speed of light.

Energy Content

α-particles are emitted with an energy in the range 4–8 MeV. Perhaps the most significant feature of the emission is that for a particular isotope the α-energy is discreet and emerges at one or a few defined energy levels. Radium-221 for instance, emits an α-particle with an energy of 6·7 MeV.

$$^{221}Ra \rightarrow \alpha\text{-particles } 6\cdot7 \text{ MeV}$$

Perhaps the most striking feature of β-particles is that they are emitted over a continuous energy range up to a maximum value. Thus, it is possible to plot the energy of the particles versus the number of them with a particular energy and this has been done for ^{3}H, ^{14}C and ^{32}P in Figure 2. Some important features emerge from these plots. Different isotopes obviously give graphs of different shape, that is, their β-energy spectra differ. It is possible to derive a mean energy from the graph which is usually about 30% of the

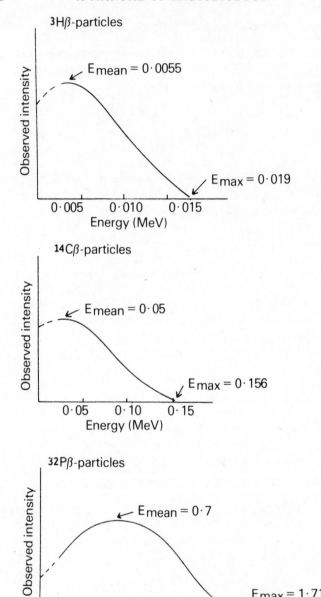

Figure 2 Beta energy curves.

maximum energy. In any particular disintegration, the energy difference between that of the β-particle emitted and the maximum energy it may have, is carried away from the nucleus by a neutrino emitted at the same time. From the graph a maximum energy is also discernible, for example

$$^{3}\text{H } E_{max} = 0\cdot019\,\text{MeV}$$

$$^{14}\text{C } E_{max} = 0\cdot156\,\text{MeV}$$

$$^{32}\text{P } E_{max} = 1\cdot710\,\text{MeV}$$

Thus, the β-particles of tritium are not very energetic, those of ^{14}C are about eight times more energetic, while those of ^{32}P are about eleven times more energetic still.

γ-radiations, because they have a short wavelength, are extremely energetic and are similar to α-particles in that they are emitted at discreet energy values, and although these are normally in the range 10 KeV to 3 MeV some have energies up to 7 MeV. Isotopes may emit γ-rays at more than one energy value.

It is well to bear in mind that the disintegration of a particular atom may release a number of radiations of the same or different types. Thus ^{22}Na emits both a β-particle and γ-radiation and ^{59}Fe emits three β-particles of different energy values.

Ionisation Effects

Because α-particles are relatively massive and doubly charged, when they pass close to another atom they may strip some electrons from this atom thus producing positive ions (see Figure 3). An α-particle with an energy of $6\cdot8$ MeV can produce about 2×10^5 positive ions in air before its energy is expended, and such ionisation ability represents a serious biological hazard.

β-particles may also lead to ionisation of materials through which they pass. However, β-particles are much less massive than α-particles and travel relatively quickly. Hence, they spend correspondingly less time in the vicinity of other atoms and have less time to remove electrons from them. The ionisation effects caused by β-particles are therefore much less intense than those produced by α-particles.

γ-rays are very different in their ionisation potential. Because the rays are uncharged they have no appreciable force fields surrounding them and can travel through matter without interacting with it to any great extent. Therefore, they do not lead directly to the ionisation of materials. However, it is of course possible for γ-rays to interact with matter to some extent and the three important ways by which this occurs are shown in Figure 3. In all cases electrons are emitted and these can cause the formation of other ions because they are so energetic. The three types of interaction result from γ-radiation of different energy levels. Low energy γ-radiation can interact directly with inner orbital electrons of atoms in the absorbing medium and, after transferring all their energy, eject an electron (termed a photoelectron). X-rays are often produced during the subsequent rearrangement of the atom. Medium energy γ-rays may interact with electrons by elastic collision and transfer a portion of their energy in the process. The electron ejected (now termed a Compton electron) changes its rate and direction of movement and can be used as a basis for measurement of γ-radiation. High energy γ-rays are energetic enough to react directly with an atomic nucleus producing both a negative and positive electron, the latter being annihilated and producing more radiation.

The consequence of these ionisation effects on auto-radiography film is discussed in Chapter 3.

Range

Because α-particles are emitted with discreet energies they will travel for a definite distance through a particular medium. Normally the distance travelled through air is measured by determining the intensity of the radiation at varying distances from the source. The results can be expressed graphically as shown in Figure 4. The range will obviously not be the same for all materials and some ranges for particles with an energy of 7 MeV are given below.

Air	Water	Al	Mica	Cu	Pb
$57,000 \mu m$	$74 \mu m$	$34 \mu m$	$29 \mu m$	$14 \mu m$	$2 \mu m$

Ionisation effects of α and β-particles

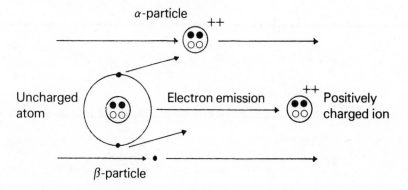

Ionisation effects of γ-radiation

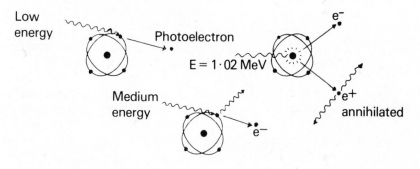

Images produced in photographic emulsions

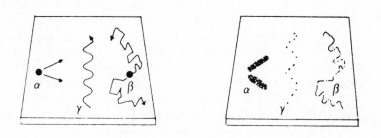

Figure 3. Ionisation effects.

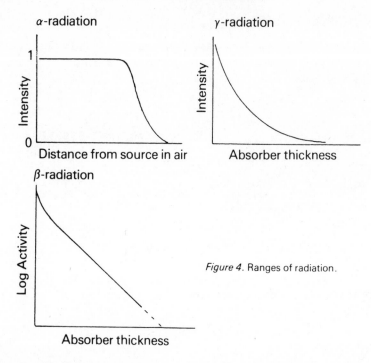

Figure 4. Ranges of radiation.

Lead will therefore form a good shielding material for these particles.

The distance β-particles travel in air will again depend upon their energy. Because β-particles do not travel in straight lines the *linear* range is difficult to measure and so the range is normally expressed in terms of the thickness of aluminium film necessary to absorb the particles. More importantly, the 'soft' β-particles from ^{3}H and ^{14}C will penetrate only short distances in air, even shorter distances in liquids and will not penetrate the walls of laboratory glassware. The 'hard' β-particles from ^{32}P will travel much further in air (about 6 m for the most energetic), shorter distances in liquids but the most energetic of them will penetrate the walls of laboratory glassware. For this reason, ^{32}P is transported in small lead containers.

As mentioned above, γ-radiation may pass through materials with little interaction, its absorption is exponential (Figure 4), hence there is no clear-cut range. Again the more energetic the radiation, the further it will penetrate an object and this is why γ-radiations present a serious biological hazard.

The properties of these ionising radiations are summarised in Appendix 1.

CHAPTER 3

COUNTING RADIOACTIVE ISOTOPES

Bearing in mind that one normally uses radioactive isotopes in the presence of non-radioactive nuclides of the same element, and that one uses very small quantities of the isotope, it is obvious that their detection by standard methods of elemental analysis is not possible.

Radioactive isotopes have two distinct features, their difference in mass from the other nuclides of that element and the fact that they emit radiation upon disintegration; both of these attributes have been discussed in the previous chapters and can be used as bases for detection methods.

The mass differences of the nuclides can be measured by mass spectrometry but this is rarely used for radioactive isotopes since there are better methods and mass spectrometers are expensive instruments and not easy to operate. However, this type of detection must be used for the non-radioactive isotopes.

With radioactive isotopes it is in fact the radiation emitted by the disintegrating nucleus that is usually detected in order to qualitatively identify and quantitatively estimate them. When a radioactive nucleus disintegrates the emission products possess enough energy to cause some change in the surrounding matter. Counting techniques and apparatus have been developed to make use of these changes in the production of electrical signals and the number of these signals produced can easily be determined.

The following discussion of counting techniques is largely

restricted to β-emitters because of their overwhelming importance in biological work compared with the α and γ-emitters.

Geiger-Muller Counting

In this type of counting the electrical signal is produced by a special tube, the structure of which is shown in Figure 5. It consists of a chamber, the inner surface being coated with an electrical conductor, and this is made the cathode of the tube. At the central axis of the chamber is a wire that is made the anode and this is insulated from the cathode.

In the most commonly used type of tube, termed an end-window tube, the end of the tube is covered with a thin membrane or window, which is permeable to particles of sufficient energy. The inside of the chamber is filled with a monatomic gas, usually argon or helium, containing a small amount (about 0·1%) of organic or halogen gas.

A number of other types of Geiger-Muller tube are also available and some of these are shown in Figure 6. The end-window type is used for most quantitative estimations on solid materials, while the thin walled tube is found mainly in contamination monitors. The remaining four types are often used to measure the activity of liquids, with the probe type being very useful for small volumes, precise localisation of radioactive materials or even for measuring radioactivity inside plant and animal tissues.

For accurate quantitative work, Geiger-Muller tubes are normally contained in a lead block or 'castle' which also surrounds the sample chamber. A typical arrangement is shown in Figure 7. The lead serves to shield the tube and chamber from outside radiations which give rise to counts in the apparatus when no sample is present. This still occurs to some extent even with the shielding and gives rise to a so-called background count rate.

When a β-emitter of sufficient energy is brought close to the window of the tube, some of the β-particles will penetrate the window and pass into the gas inside the tube leading to the formation of pairs of ions, usually positive ions and electrons. If a high potential difference is applied across the

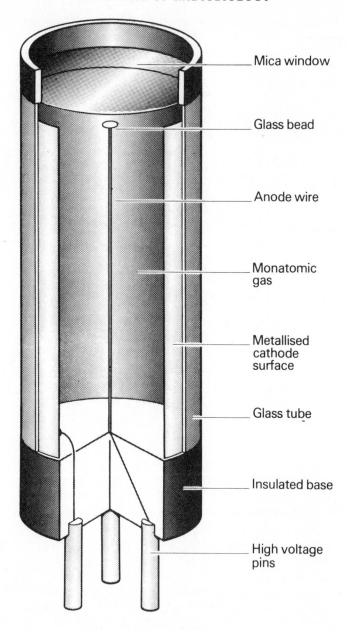

Mica window

Glass bead

Anode wire

Monatomic gas

Metallised cathode surface

Glass tube

Insulated base

High voltage pins

Figure 5. Geiger-Muller tube.

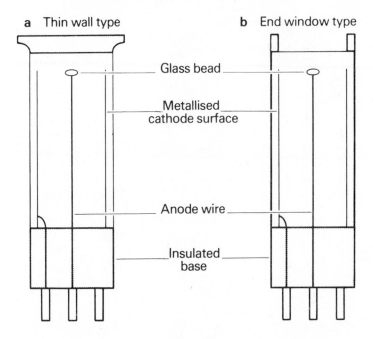

Figure 6 (above and on page 22). Types of Geiger-Muller tubes.

electrodes these ions are accelerated towards the electrode of opposite charge. The accelerated ions also react with the gas atoms in the tube to produce more ions and this chain reaction continues to produce an avalanche of ions. An amplification of about 10^6 to 10^8 is normally obtained. When the ion avalanche reaches the electrodes it is neutralised producing a flow of electrons in the external circuit and giving a measurable potential of one to ten volts.

If the avalanche producing chain reaction was not terminated immediately on contact with the electrode, it would continue for some time and during this time the tube could not detect another β-particle. Hence the inclusion of the organic or halogen gas to absorb some of the energy of the accelerated ions and thus *quench* them, thereby terminating the reaction after a very short time. Sufficient quenching gas must be added to return the tube to its unionised state in less than a millisecond from β-particle entry.

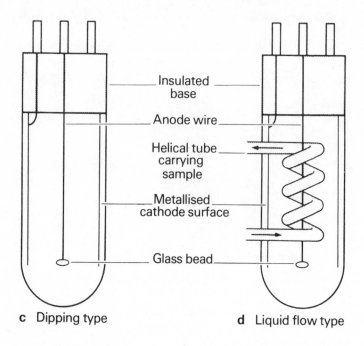

c Dipping type   **d** Liquid flow type

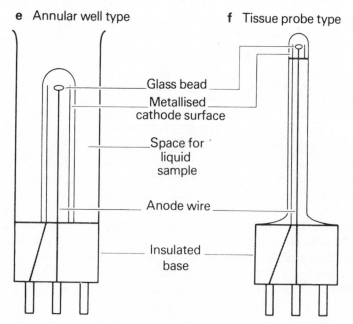

e Annular well type   **f** Tissue probe type

Figure 6 continued

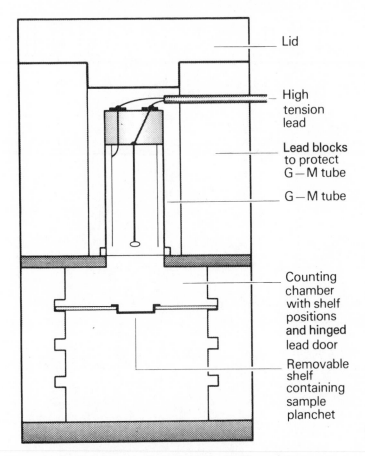

Figure 7. Diagram of a typical Geiger-Muller apparatus.

The electronic circuits associated with the tube can be designed to indicate the average current flow (a rate meter) or to total the number of electrical pulses or 'counts' (a scaler). The count rate is dependent upon the potential applied across the electrodes and this can be shown by varying the applied voltage and determining the count rate on a suitable sample. The results of such an experiment can be expressed graphically as shown in Figure 8. At low voltages the curve is exponential and slight changes in voltage cause considerable changes in count rate. At higher

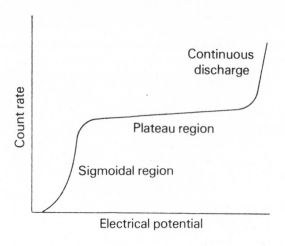

Figure 8. Geiger-Muller detector curve.

voltages the curve becomes almost linear and horizontal, in fact it has about a 5% rise, and this is termed the plateau region of the tube. Only over this region is the voltage across the electrodes great enough to cause particles to move sufficiently fast to produce maximum ionisation of the tube gas. The presence of this plateau region is useful in that expensive, stabilised high voltage power supplies are not required. At even higher applied voltages, the count rate increases exponentially as the Geiger tube goes into continuous discharge, the voltage being great enough to break down the insulating properties of the gas inside the tube. For maximum stability the tube is operated in the plateau region where changes in voltage have little effect upon count rate. To prolong the life of a tube it is generally operated at a voltage of only about one-third the plateau length above the low voltage end of the plateau.

In the plateau region the tube is operating at near its maximum efficiency and therefore no distinction can be made between particles of differing energy content, providing of course the particles have sufficient energy to penetrate the tube window in the first place. This method of counting is therefore an 'all or nothing' process.

The efficiency of the tube will be given by:

$$\frac{\text{counts per second from the sample}}{\text{disintegrations per second from the radioisotope}} \times 100$$

When the energy of the β-particle is low, for example from tritium, most of the particles are absorbed either before they reach the Geiger tube of the counter or by the window material. Hence, tritium cannot be counted on this type of system unless special windowless gas flow tubes are used. The β-energies of ^{14}C and ^{32}P are sufficient to enable these radioisotopes to be counted on end-window systems though the efficiency is still low.

In order to increase the efficiency of counting, tubes with extremely thin windows are sometimes used. Since small amounts of gas can leak through such a tube window, a continuous flow of gas is required. These tubes are of course only suitable for quantitative experimental work and not for portable contamination monitors.

The conditions under which a Geiger-Muller tube is usually operated are such that the tube gas is almost totally ionised and hence electrical pulse size is not influenced by the nature of the incident radiation. However, at lower voltages α- and β-particles produce different degrees of ionisation and different intensities of signal. Consequently, by use of suitable discriminator circuits, either in a normal Geiger system or in a low voltage ionisation chamber, α-particles may be detected and even α- and β-radiation from the same sample counted separately. This type of counting is termed proportional counting since the size of the signal is proportional to the amount of ionisation produced by the incident radiation and hence varies with the nature of this radiation. Because of the marked effect of slight changes in applied voltage, this type of counting system is not as satisfactory as more normal Geiger-Muller counting.

Another problem is that because of the ready absorption of α-radiation by materials, an extremely thin window in the counting tube is required in order that a high proportion of the radiation will pass through.

Scintillation Counting

With the introduction of commercial scintillation counters in the mid-1950's, radiobiology and biology in general took a considerable step forward. This is partly because of the ease of sample preparation but principally because of the high counting efficiencies obtained with all radioisotopes. The ability to count tritium radiation is particularly important since most natural compounds can be quite easily, and hence cheaply, labelled with tritium and, since tritium is carrier free, the labelled compounds have very high specific activities.

Although scintillation counting depends upon the release of ionising radiations from an isotope it is not ionisation which gives rise to the electrical pulse, as in a Geiger-Muller counter. Instead the energy of the radiations is used to produce pulses of light and it is these pulses which are counted. The conversion is carried out by substances usually called *fluors* or *scintillators* because they fluoresce upon irradiation, and since fluorescence is an important aspect of scintillation counting, we will briefly review its mechanism here.

The MECHANISM OF FLUORESCENCE

According to the modern theories of atomic structure, electrons are contained in shells or energy levels surrounding the nucleus. It has been found that each of these major energy levels can be subdivided into a number of subshells which differ slightly in their energy content. Under normal conditions, the electrons will be found to occupy the sublevel with the lowest energy for the greatest proportion of the time. This is termed the *ground-state* of the atom. However, it is possible for atoms to absorb energy ($h\gamma_1$), the electrons then passing from the ground state to a higher or excited level (see Figure 9). For this to occur the energy of the quantum absorbed must equal exactly the difference in energy between the two sublevels. In the excited state the atom is less stable than in the ground state and will decay back to the latter state releasing the absorbed energy in the process. If this energy is released all at once, then the energy

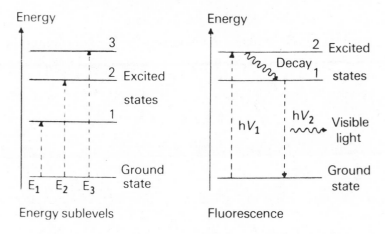

Figure 9. Electronic sublevels and fluorescence.

of the released quantum will be the same as that absorbed and so the energy emitted will have the same wavelength as that originally absorbed. Thus, if the compound absorbed ultraviolet light of a particular wavelength, ultraviolet light of the same wavelength would be re-emitted. However, this does not always occur. It is possible for the excited state to decay to a sublevel of only slightly lower energy by converting some of the energy associated with the excited electrons into other forms. If now the remaining energy is liberated all at once, the energy of this quantum ($h\nu_2$) will be lower and hence the radiation will have a longer wavelength and may appear in the visible region of the spectrum. This latter situation occurs in fluorescence; many scintillators, for example, when irradiated with ultraviolet light (short wavelength) emitting visible blue light (longer wavelength).

During scintillation counting some of the energy of the radiation causes the scintillator to fluoresce in the visible region of the spectrum and these fluorescence pulses traverse the transparent scintillator and can be counted by the associated electronic equipment. It is extremely valuable that the number of light photons released is proportional to the radiation energy absorbed (about seven photons per keV of energy in a liquid scintillator) and hence differentiation between radioisotopes emitting β-particles of different energies is sometimes possible. A particular β-particle is said to produce a particular pulse height, because its energy yields a light pulse of a particular size or number of photons. Owing to the nature of β-emissions, a β-emitting radioisotope will yield a range of pulse height originating from β-particles of different energies.

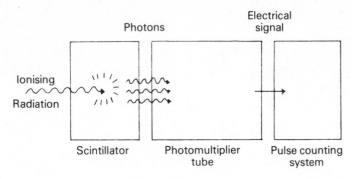

Figure 10. Simplified mechanism of scintillation counting

SCINTILLATION COUNTING SYSTEMS

The process of scintillation counting is shown very simply in Figure 10. The radiation is absorbed by the scintillator and this responds by fluorescing and giving out photons of visible light. These photons are captured by a photomultiplier tube which as a consequence produces a large number of electrons in response to the capture.

Figure 11 shows the construction of a typical photo-multiplier tube with a solid scintillator attached. The function of the photomultiplier is to generate, from the fluorescent light photon, an electrical signal and to amplify it to a measurable level. In order to achieve maximum efficiency in this it is important that the absorption characteristics match reasonably well the spectral output of the scintillator. If this is the case the photons captured by the tube interact with its photocathode and this responds by emitting electrons. These electrons are then attracted to a positive electrode often called a dynode, and this produces three to six electrons in response. These electrons are then attracted to another dynode which is held at a slightly higher potential than the first and this again emits more electrons. The process is repeated along the length of the tube and a measurable current is collected at the end. Photomultipliers are available with six to fourteen dynodes and the magnification obtained is considerable. About 10^7 electrons arrive at the collector for each electron produced by the cathode. However, photomultiplier tubes suffer from the problem that they are sensitive to external magnetic fields and to temperature, and that they require a stabilised voltage supply.

There are two types of scintillator:

(i) inorganic scintillators which generally consist of a sodium iodide crystal activated with thallium (for α-radiation) or zinc sulphide (for γ-radiation); and

(ii) organic scintillators, in which single crystals, plastic discs, rods or cylinders and solutions of organic compounds, are used.

Scintillation counters themselves can also be divided into two major types depending upon the relationship of the sample to the scintillator. External sample counters have the

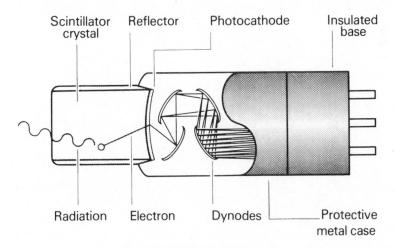

Figure 11. Photomultiplier tube

sample and the scintillator in close contact and this is usually ensured by placing the sample container next to the scintillator. Internal sample counters on the other hand, have the sample and the scintillator intimately mixed in a solution or suspension.

We will now consider each of these in turn.

EXTERNAL SCINTILLATION SYSTEMS

This type of counting can be used for solid materials, solutions of solids and for liquids. The material to be counted is placed in a suitable container and put in close contact with the scintillator. A typical piece of apparatus for achieving this is shown in Figure 12. The apparatus is designed to accept a test tube containing the sample, though other types, including flow-through cells, are available. Essentially, it consists of a lead container into the base of which is built a photomultiplier tube. The sample tube fits into a well which has a movable light-tight lid and which is situated in the top of the apparatus. In the simplest

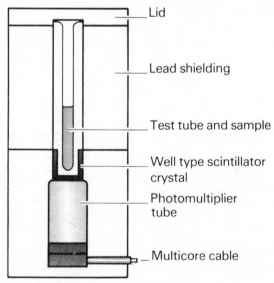

Lid

Lead shielding

Test tube and sample

Well type scintillator crystal

Photomultiplier tube

Multicore cable

Figure 12. Typical solid external scintillation counter.

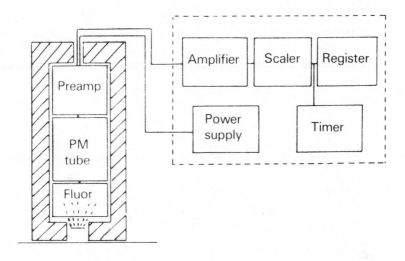

Figure 13. Layout of an external scintillation system.

arrangement the output from the photomultiplier tube is passed via an electronic amplifier system to a normal scaler unit which will provide a numerical indication of the radiation emitted by the sample (Figure 13).

A significant problem with solid scintillation counting is that because the samples are placed in a container of some type for counting and this is separate from the scintillator, the radiations must still be able to penetrate through the container and pass into the scintillator. Hence in general this type of counter can only be used for those isotopes emitting relatively energetic radiations; tritium, for example, cannot be counted by this method and this deficiency has lead to the technique being supplanted by liquid scintillation counting.

LIQUID INTERNAL SCINTILLATION COUNTING

In this type of scintillation counting an organic scintillator is dissolved in an organic solvent and the sample to be counted is dissolved in this mixture. In this type of system the compound to be counted, the solvent molecules and the

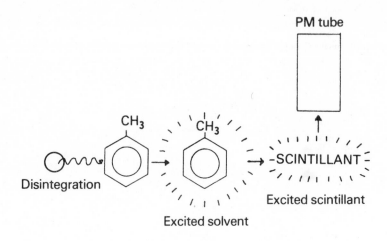

Figure 14. Simple mechanism of scintillation counting.

scintillator molecules are all in intimate contact and hence a much more efficient transfer of radiation energy is obtained than is possible with external scintillation counting.

In simple terms, some of the energy of the radioactive disintegration is captured by the solvent molecules which are thus raised to a higher energy content, that is, they become excited. If these excited molecules simply released the energy again this would be of little use, but in fact a reasonable proportion of them transfer the energy to the scintillator molecules which in turn fluoresce, giving out photons which may be captured by the photomultiplier tubes. This very much simplified scheme is shown in Figure 14.

A number of organic scintillators have found widespread use, three of the more important being *p*-terphenyl, 2,5-diphenyloxazole (PPO) and 1,4-bis-2-(5-phenyloxazolyl)-benzene (POPOP). In the simple mechanism depicted in Figure 14 a single scintillator was used. However, in some cases, the wavelength of light emitted by the scintillator may not match that most efficiently detected by the photomultiplier tubes. Hence, many scintillation fluids contain a *secondary* scintillator (or wavelength shifter) which absorbs the energy from the *primary* scintillator and re-emits

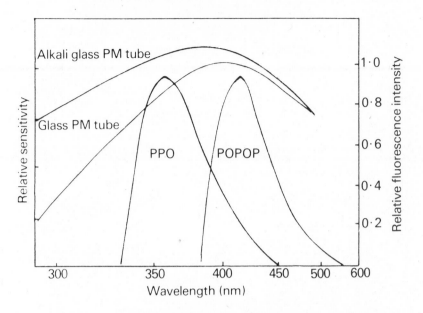

Figure 15. Spectral sensitivity of PM tubes.

it as light of longer wavelength. *p*-terphenyl and PPO act as primary scintillators and POPOP as a secondary one. The process is depicted graphically in figure 15. Modern developments in both the design of photomultiplier tubes and the production of new scintillants is leading to a situation where the need for such secondary scintillators might be unnecessary.

Figure 16 depicts diagrammatically the components of a simple liquid scintillation counter. The solution of the sample is contained in a glass sample tube or bottle and is placed between two photomultiplier tubes. The latter produce momentary current flows in response to each scintillation and these currents are amplified and applied to a rate meter, scaler or more sophisticated computational facilities. Scintillation counters normally contain a co-incidence unit which allows only scintillations detected by both tubes simultaneously to be counted and in this way the serious errors introduced by counting the spurious electrons (thermal electrons) generated by the photomultiplier tubes and by counting cosmic radiation can be minimised.

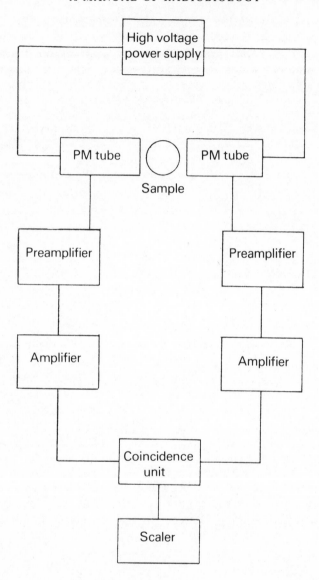

Figure 16. Block diagram of a simple scintillation counter.

The intensity of the scintillation, and hence the current pulse produced, depends upon the energy of the β-particle producing it. Many scintillation counters contain a *pulse height analyser* which accepts for counting only scintillations of certain energy intensities (pulse heights). The intensities to be counted may be selected to lie between two limits and this 'window' can be varied over an arbitrary voltage range by the operator. This makes it possible to distinguish between two or three radioisotopes in the same sample providing their maximum β-energies differ by a factor of five or more (e.g. ^{3}H, ^{14}C and ^{32}P).

Sophisticated scintillation counters capable of double isotope measurement are usually multichannel instruments with two or three independent analysis channels comprising amplifiers, pulse height analysers and scalers. Each channel can be selected to count one of the required isotopes and the problems of overlap of the β-spectra of certain isotopes can be partly resolved by the calculation equipment associated with the instrument.

Cerenkov Counting

It is interesting that β-particles of sufficiently high energy cause the production of weak ultraviolet light when passing through water. This so-called Cerenkov radiation can be counted directly by scintillation counters with the advantage that quenching is severely reduced and expensive scintillants and solvents are not required. While ^{14}C cannot be detected in this way ^{32}P can be detected with an efficiency of about 40%.

Gamma Radiation Counting

Gamma-emitting samples are often estimated in external scintillation systems using well-type sodium iodide crystals since normal liquid scintillation cocktails have poor response to γ-radiation. However, special (and expensive) sealed vials are now available containing a tin-loaded scintillant which will respond to γ-radiation. The sample is placed in a sealed plastic tube and inserted into a well in the scintillation vial, thus enabling the vial to be re-used.

Counting Errors and Corrections

There are a number of sources of error in all types of counting which will lead to inaccurate results and while some of these have been alluded to previously a more complete discussion of them and the available corrections is necessary.

These errors can be grouped into two categories; those concerned with the sample and those associated with the counting instrument and procedures.

SAMPLE-BASED ERRORS

The problem of sample geometry concerns the shape of the sample and its distance from, and orientation to, the detector. It is particularly important for Geiger-Muller counters since particles frequently fail to penetrate the tube because they are absorbed within the sample itself (self-absorption) due to its thickness or to the weakness of the emission. They may also be absorbed in the air between the sample and the detector or by the detector window. The highest counting efficiency will be achieved when the sample is spread evenly over a flat surface and there is only a little of it. This is known as infinitely thin plating of a sample and it allows β-particles to leave the film with little self-absorption and indeed particles initially travelling away from the detector may be reflected towards it by the planchet itself. It is also possible to use very thick layers of material and this is known as infinitely thick plating of samples. Only emissions close to the surface will leave the film and be detected but more reproducible plating of samples, particularly suspensions, can be obtained.

The distance of the sample from the detector window is also of importance. The nearer the sample is to the window the smaller will be the proportion of particles absorbed in the intervening air and the greater the proportion of particles emitted by the sample which will be initially directed towards the window.

In many tracer studies one is more concerned with the relative activities of various samples than with the absolute rate of disintegration, and numerical corrections are not often necessary providing the samples are counted on the same size planchet, at the same distance from the tube, for the same time and contain approximately the same weight of sample.

The efficiency of scintillation counting can also be impaired by these problems although most of them are not significant; the nature of the apparatus ensuring optimum and reproducible sample placement, variation in sample volume having little effect on counting efficiency and in addition the reflective nature of the sample chamber ensuring good light trapping by the photomultipliers. However, self-absorption (usually known as quenching) is an important problem meriting further discussion.

Quenching is a term given to the processes resulting in a reduction of the amount of light reaching the photo-multiplier and it constitutes the most serious problem in the scintillation counting of biological samples. Quenching may be due to interference with the transfer of energy from the β-particle through the solvent/scintillator system to the photomultiplier (chemical quenching) or to the absorption of light in the scintillation fluid (colour quenching). Thus many coloured materials are difficult and sometimes impossible to measure efficiently by scintillation counting unless bleached, because of their absorption of light photons emitted by the scintillators. Other materials causing serious quenching are strong acids, halogenated compounds and dissolved oxygen. It is also worth remembering that dirty vials and turbid solutions may also cause a reduction in light emission, a process termed optical quenching. The main effects of quenching are to reduce the pulse intensities and to shift the β-spectrum to lower energies. One can foresee the decrease in counting efficiency and the introduction of variable efficiencies for different samples that this would cause.

INSTRUMENT AND PROCEDURAL-BASED ERRORS

One has to bear in mind the reliability of the count rate figures produced by the instrument. The time at which a given nucleus will disintegrate does not depend upon its previous history or on its surrounding environment; it is an entirely random process. Repeated count rates determined on the same sample are therefore unlikely to be equal but will fluctuate about a mean value and ideally one should not merely count samples for an arbitrarily and possibly conveniently chosen time interval, but until a statistically reliable number of counts have been obtained. For example, if only 100 counts were obtained there will be a one in three chance of the error exceeding 10%, whereas for 1,000 counts this error drops to $1 \cdot 16\%$. It is worth noting, however, that although very active samples need to be counted for only short periods the paralysis time of the Geiger-Muller tube and associated electronics may become very significant and may need to be taken into account.

The mode of action of the Geiger-Muller tube means that there is a short time lapse after a particle has entered a counting tube during which the tube cannot respond to another particle. This is the paralysis or resolving time and it is made up of a period of dead time and recovery time. The errors due to this become significant as the count rate rises (about $0 \cdot 1\%/300$ counts s^{-1}) and tables are available giving the appropriate corrections for different count rates assuming a particular resolving time (see Appendix 2).

The problem of paralysis does not normally occur in scintillation counting, since the scintillants recover within 10 ns and the electronics can process pulses within 100 ns but modern instruments nevertheless take account of this.

A final problem is that a low count rate will be observed on all instruments even when no radioactive sample is present. This 'background radiation' arises, in Geiger-Muller counting, mainly from cosmic radiation and radioisotopes present in the materials used in the construction of the instrument. It can also arise from contamination of the counter or from radioactive isotopes in the near vicinity. For example, $^{32}PO_4^{3-}$ has a tendency to adhere to glass and should never be counted in any of the liquid Geiger-Muller tubes described earlier. Counting apparatus should not normally be situated in a laboratory used for radioactive experiments or for isotope storage.

Another problem with older types of scintillation solution

is that they suffer from the tendency to scintillate due to the absorption of heat energy at room temperature and also from absorption of light energy. These problems may be overcome by keeping the samples cold and in the dark for at least one hour before counting or by using one of the newer (but more costly) scintillation fluids.

In scintillation counters thermal noise in the photomultiplier units contributes to the background count but this can be reduced by cooling the tubes and by using coincidence circuitry and pulse height analysers as described previously.

Background counts may also be reduced by shielding the instrument usually with lead encasements and some manufacturers go to the length of obtaining metals made before the discovery of atomic fission in an attempt to reduce the background radiation produced from within their counters. The problem of background radiation is that because it is a random process repeated counts are not of the same value and so the reliability of low experimental count rates is poor.

STANDARDS

The problems discussed above, with the exception of background radiation, can be largely overcome by the maintenance of reproducible geometry and instrument conditions, and by the use of standards.

If a planchet is plated with an infinitely thin or an infinitely thick film of a standard compound of known activity, this can be used to determine the counting efficiency of the instrument by observing the reduction in the count rate below the theoretical or actual value of the standard. Having determined the efficiency a simple calculation enables one to find the absolute activity of an experimental sample. With scintillation counting it is unfortunate that because of the variation in sample preparation and the degree of quenching the counting efficiency must be determined on each sample using either internal or external standards. Internal standards are stable materials of well defined composition (such as ^{14}C-n-hexadecane), a known amount of which can be added to the sample and the counting efficiency determined by calculating the reduction in count rate produced by the sample. This method is time-consuming, subject to pipetting errors and ruins the original sample. It is, however, probably the most accurate method.

Modern scintillation counters contain a γ-ray emitting isotope which can be moved to be adjacent to the sample vial. Again the reduction in count rate produced by the sample can be used to calculate the counting efficiency. With more sophisticated multichannel scintillation counters methods based on the ratios of counts produced in two of the instrument channels are used.

Autoradiography

Autoradiography is the detection of radioactivity by using photographic emulsions. A photographic emulsion consists of very small crystals of silver bromide in gelatine mounted on a glass or flexible support. However, the silver bromide is not pure but also contains small quantities of silver sulphide and colloidal silver. Also the lattice structure is not perfect but contains spaces or 'holes' in the structure where ions should be and there are silver ions out of place and in between the other ions (interstitial ions).

When a radioactive emission passes into a photographic emulsion it causes ionisation and sets electrons free in the ionic silver bromide lattice. The electrons migrate over short distances to areas known as 'sensitivity specks' which are probably composed of silver sulphide and these in turn become negatively charged. This charge attracts the interstitial silver ions and upon reaching the speck they are neutralised to metallic silver forming a so-called latent image. Thus, the specks act as loci for the growth of nuclei of metallic silver, the amount of metallic silver formed at these points being dependent upon the amount of radiation received.

During development of the film, the developer supplies electrons and causes the reduction of silver ions to metallic silver atoms and these are responsible for the visible image. However, because of the catalytic effect of the metallic atoms already produced by the radiation, the silver ions in the region of the latent image are reduced by the developer very much faster than those in the regions where there is no latent image. Hence, if development is correctly timed, only the former produce a visible image and this image will reflect the sites of exposure of the emulsion to radiation. Thiosulphate present in the fixer solution removes unexposed silver bromide and hence stabilises the image.

The technique of autoradiography consists of placing a radioactive specimen next to a photographic emulsion and locating the sites of activity within the specimen by examination of the image produced on the emulsion. However, because one speck of silver initially present only grows by one silver atom for each quantum of energy absorbed, weak radioactive sources will require long exposure times to give a good visible image. The length of exposure time can be calculated to give about 10^5 β-particles mm^{-2} of emulsion or it can be determined by trial and error. An advantage of the technique is that the length of exposure is only limited by the decay of the isotope and latent image and by fogging of the emulsion. It is possible therefore to detect very small quantities of isotope by prolonged exposure.

Against this there are the disadvantages that poor resolution (i.e. image quality) may occur because of the nature of the emulsion or due to movement of the specimen by shrinkage or for other reasons. Realignment of the specimen for isotope location is sometimes difficult, there is a delay in obtaining the experimental results and it is difficult to make the technique quantitative.

It is obvious that intimate contact between the film emulsion and the radioactive source is essential in order to get maximum transfer of energy from sample to film. Hence, paper and thin layer chromatograms are generally laid straight on to the emulsion and held in contact by a heavy weight. Whole plants are usually dried and pressed and animals frozen and sliced with a band saw before laying on to the film.

It is possible to carry out autoradiographic localisation of isotopes using histological material by permanently covering the tissue section or cell smear on a slide or electron microscope grid with emulsion. After exposure and development the material is viewed in the normal way and the presence of isotope is apparent as dark areas overlying the biological material. There is no need to realign specimen and film and the fine particle size of the emulsion used produces very sharp images. The emulsion can be applied to the specimen either by using the specimen to capture a piece of nuclear stripping film floating on water, by dipping the specimen into a bath of liquid emulsion or by placing the specimen on a slide already coated with emulsion.

The amount of ionisation produced by an emission will have a bearing on the intensity and sharpness of the photographic image. The α-particle is relatively heavy, has a short path length and will produce ionisation throughout its path length. γ-rays on the other hand are very energetic and can pass through emulsions causing little ionisation and poor images. The β-particle is lighter, produces less ionisation per unit length and travels longer distances than does the α-particle if both have the same initial energy. The β-particle is relatively easily deflected from its path by heavy atoms, its path is random compared with the short, straight course of the heavy α-particles and this leads to diffuse images. The quality of the image is also dependent on the path length and hence the energy of the β-particle. Thus ^{3}H with a path length of 1 μm produces very good images, ^{32}P with a path length of 800 μm produces poor ones. These effects are illustrated diagrammatically in Figure 3.

Since a point source emits particles equally in all directions, the particles will strike the emulsion within a hemisphere, with the radius being the range of the particles from the point source. Other considerations aside, the lower the particle energy, the thinner the specimen, the closer the contact between specimen and emulsion and the thinner the emulsion, the better will be the resolution of the image. The X-ray films without a lead-based intensifying screen, which are usually used for macroscopic work, have large silver bromide grains of varying size and thus give poor resolution in terms of microscopic images. Nuclear stripping films have the best resolution for microscopic images because of the small uniform grains and a greater concentration of grains per unit area. X-ray films containing lead screens are not suitable for autoradiography using ^{14}C and ^{3}H because of the low penetrating power of the emissions produced by these isotopes. Some photographic emulsions commonly used in the autoradiography of biological specimens are listed in Appendix 4.

The degree of fogging of emulsions due to background radiation increases with age and this will decrease the usefulness of the film. Thus, films for autoradiography must be stored well away from radioactive sources and should not be stored for too long before use.

RADIOCHROMATOGRAMS

Autoradiography has been extensively used for the localisation and identification of radioactive materials on paper and thin layer chromatograms. However, the lack of quantitative data from the technique is a serious disadvantage and a variety of procedures are used to overcome this. Commercial or home-made scanners can survey the chromatogram using a windowless gas-flow Geiger system and indicate intensity and localisation of activity on a chart recorder. Alternatively, a paper chromatogram can be cut into sections or a thin layer chromatogram eluted and the activity assessed in a scintillation counter or in a Geiger-Muller system.

CHAPTER 4

BIOLOGICAL USES OF RADIOISOTOPES

The usefulness of radioisotopes in biological studies stems largely from the following facts. The methods used to count isotopes are extremely sensitive and therefore very small quantities can be detected in large quantities of unlabelled material. An activity of as little as 10^{-8} to 10^{-10} Curies can be detected by ordinary equipment although the quantity of material corresponding to this activity depends upon the specific activity of the sample, and the quantity of radioactive material on the half-life of the isotope. A comparison of this limit of measurement with other methods of analysis of for example, phosphorus, shows the great advantage of radiochemical methods.

Gravimetric analysis	10^{-6}g
Spectrophotometry	10^{-7}g
Emission spectroscopy	10^{-8}g
Polarography	10^{-10}g
Radiochemical analysis	10^{-16}g

The considerable sensitivity of the available detection methods enables one to use very small amounts of radioactive material and this will mean that the cost of the experiments is often kept low. It is also of some importance in many physiological experiments where the normal metabolic balance could easily be upset by unnaturally high concentrations of material added merely to facilitate detection. Detection and estimation of isotopes by monitoring their radiation also has the advantage that the sample is not always destroyed or altered and may be retained for further study.

With the exception of some of the lighter elements, radioactive and non-radioactive nuclides of the same element behave in an identical manner in chemical reactions. Hence, compounds containing radioactive isotopes will be metabolised in the same way as those that do not and one can be certain of looking at the true biological process. Also, because radioactive and non-radioactive samples of the same compound are chemically equivalent, small amounts of labelled compounds become intimately mixed with much larger quantities of unlabelled material in the organism and cannot be distinguished from it except for their radioactivity. Therefore, isotopically labelled compounds can be detected in the presence of pre-existing unlabelled material and, as discussed later, this is of great significance in many studies.

Another important point which has been mentioned in an earlier chapter but could be reiterated here is that the properties of the isotope do not vary with the compound into which it is incorporated. Thus the detection and estimation of an isotope, in contrast to most chemical and instrumental measurement techniques, is not influenced by the chemical nature of the sample.

Some of the main ways in which isotopes have been used in biological experimentation will now be discussed by taking examples from investigations in biological systems of different complexity.

Isotope Dilution

This technique is particularly useful when it is extremely difficult to separate and isolate all of a compound for quantitative estimation. If a radioactive specimen of one of the compounds present in the mixture is added to the mixture, the label becomes uniformly distributed throughout the mixture. When the compound is then re-isolated some of the label will be present and hence can be counted and the specific activity of the sample calculated. The concentration of the non-radioactive compound in the mixture can be determined from the extent to which it has diluted the radioactive additive i.e.

Weight unlabelled material =

$$\text{Weight labelled} \left(\frac{\text{Specific activity additive}}{\text{Specific activity isolate}} - 1 \right)$$

It is not necessary to isolate all the material under study from the mixture because the label will be uniformly diluted in it. It is, however, necessary to purify the extract rigorously to remove any radioactive contaminants.

The following examples where the method has been of value will show the usefulness of the technique.

AMINO ACID CONTENT OF PROTEINS

Proteins consist of about 20 amino acids present in either large amounts or trace amounts. The structural determination of proteins requires the amount of each amino acid to be known accurately and this can prove extremely difficult to find. If a labelled amino acid is added to the protein hydrolysate and then re-isolated by paper chromatography, ion exchange chromatography, gel filtration or electrophoresis, the percentage of that amino acid present can be found by the isotope dilution equation.

RADIO-IMMUNO ASSAY

This is a technique that has only fairly recently been developed and because of its sensitivity it is likely to become increasingly important in the future.

The technique is perhaps most important in the clinical field for studying the level of certain hormones in the blood system. A large number of hormones are polypeptides, for example somatotrophin (growth hormone), adrenocorticotrophic hormone (ACTH) and insulin, and obviously abnormal changes in their blood concentrations are of great significance to an individual and of considerable interest to the clinician. It is very desirable to be able to determine accurately the level of hormone in the blood in order to be able to predict the development of various diseases or syndromes, to identify the cause of them and to be able to follow the effectiveness of any course of treatment. This can, however, be extremely difficult because of the very low concentrations of these hormones in the blood stream. The technique of radio-immuno assay, which is based on the isotope dilution principle, can be used to detect picogramme $(10^{-12}g)$ quantities of certain hormones.

If human proteinaceous hormones are injected into, say, a rabbit they will be treated as foreign materials and the rabbit will produce specific antibodies to them during the course of its immune response. Such blood can be used as the basis of methods for the assay of human hormones.

In order to carry out the assay it is essential that a supply of purified physiologically active hormone is available. Growth hormone, for example, is obtained by extracting post-mortem human pituitary glands and purified by standard protein purification techniques. It then has to be radioactively labelled under extremely mild conditions and for this it is treated with ^{125}I or ^{131}I in order to label the tyrosine residues in the protein. The labelled hormone is separated from unlabelled material and decomposition products by electrophoresis.

Now, if a mixture of the unlabelled hormone and labelled hormone of a known concentration is added to suitably diluted rabbit plasma containing the antibody and this is incubated at 4°C for an appropriate time, the hormone molecules will become attached to the antibody molecules. The products of this reaction are large complexes which can be collected by centrifugation and their radioactivity determined by normal counting procedures. The number of hormone molecules attached to the antibodies is fixed by the concentration of the hormone (assuming there is an excess of the antibody) and the amount of radioactive hormone attached depends upon the relative proportions of labelled and unlabelled hormone. The count rate obtained from the sample is an indication of the amount of labelled hormone attached to the antibody and a calibration curve can be obtained by increasing the proportion of labelled to unlabelled hormone.

If now a suitably diluted human blood plasma is substituted for unlabelled hormone and the complex produced is isolated and counted, then the amount of hormone in the plasma can be determined from the calibration curve with great accuracy and sensitivity.

It should be borne in mind that isotope dilution is not restricted to the estimation of chemicals but can be and indeed has been, used to measure other parameters where radioactive specimens can be produced and homogeneously mixed with the original sample. The determination of blood volume or red cell number or of insect population size is frequently made by this technique.

BLOOD VOLUME DETERMINATION

A known volume of radioactive material is added to an unknown volume of blood and the label is given time to mix thoroughly. A known volume of blood is then removed and the total blood volume found by determining the isotope dilution.

It is possible to attach the label to either the red blood cells using ^{51}Cr or to the plasma albumin using ^{125}I. Each method has its disadvantages, red cells for instance may not mix too well with the unlabelled ones and tend to travel round the circulatory system in a pulse. Albumin is distributed throughout the plasma space and part of the lymphatic system so that the volume measured may be slightly high.

Enzyme Assay and Studies of Enzyme Reaction Mechanisms

Enzyme assay techniques involving isotopes are very versatile, offer extremely high sensitivity and give accurate and reproducible results. The sensitivity of radiochemical assays enables them to be used over a wider range of substrate concentrations than is possible with other methods. This makes them most useful for the determination of Michaelis constants, for studies of competitive inhibition at low substrate and inhibitor concentrations and for assaying small amounts of, or small changes in the amount of enzymes.

For an example of this the reader is referred to the experiment later in this book where the activity of glutamate decarboxylase enzyme is monitored by measurements of the $^{14}CO_2$ released from a labelled acidic substrate. Reaction rates of a few micromoles substrate metabolised per minute can quite easily be measured.

The availability of isotopically labelled substrates has revolutionised studies of enzyme reaction mechanisms. The great sensitivity of the technique enables some enzyme-substrate complexes to he isolated and studied. However, undoubtedly the most important facet of this work is that the isotopic substrate need not be universally labelled but instead specific atoms within the molecule may be labelled enabling precise studies of chemical rearrangements in the substrate to be made.

It can be shown for example that enzymes may react with only one of two identical groups in a supposedly symmetrical molecule and that they always react with that group.

In cells, citrate is formed by a series of Krebs' cycle reactions summarised as follows:

$$CH_3.COSCoA + \underset{\underset{\text{Oxaloacetate}}{CH_2.COOH}}{\overset{CO.COOH}{|}} + H_2O \longrightarrow \underset{\underset{\text{Citrate}}{CH_2.COOH}}{\overset{\overset{CH_2.COOH}{|}}{HO.C.COOH}} + CoASH$$

$$\underset{\text{Acetyl CoA}}{}$$

It is apparent that the acetate unit forms one of the $CH_2.COOH$ groups in citrate and that citrate is a symmetrical molecule with two identical $CH_2.COOH$ groups.

Citrate is broken down to α-ketoglutarate as follows:

$$\underset{\underset{CH_2.COOH}{|}}{\overset{\overset{CH_2.COOH}{|}}{HO.C.COOH}} \longrightarrow \underset{\underset{\underset{\alpha\text{-ketoglutarate}}{CH_2.COOH}}{|}}{\overset{\overset{CO.COOH}{|}}{CH_2}} + CO_2 + 2H$$

Thus, if labelled acetyl CoA was used to produce citrate it would be expected that both the carboxyl groups of α-ketoglutarate should be equally labelled since the enzyme producing it could theoretically attack either end of the citrate molecule. In fact, analysis of the α-ketoglutarate formed reveals that the label is exclusively in the $CH_2.COOH$ grouping and therefore the enzyme responsible for the breakdown of citrate must act specifically on only one of the $CH_2.COOH$ groups and this is not the one formed from acetyl CoA. Such a result can only be achieved if there are three points of attachment between the enzyme and citrate and if the two $CH_2.COOH$ groups are not geometrically equivalent. This so-called 'Ogston effect' is shown diagrammatically below.

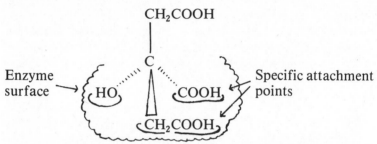

Notice that the arrangement of groups on the other three faces of the tetrahedron is different and thus there is only one way in which citrate can attach to the enzyme. This experiment would have been inconceivable without the use of radioisotopes.

There are numerous isomerisation reactions in the cell and a suitable example is that carried out by the enzyme phosphoglyceromutase which catalyses the following reaction:

$$
\begin{array}{ccc}
\text{COOH} & & \text{COOH} \\
| & & | \\
\text{CHOH} & \longrightarrow & \text{CHO} \; \boxed{P} \\
| & & | \\
\text{CH}_2\text{O} \; \boxed{P} & & \text{CH}_2\text{OH}
\end{array}
$$

3-phosphoglycerate 2-phosphoglycerate

Isotopically labelled phosphate was used to distinguish between two possible mechanisms for this reaction. In the first mechanism, phosphate may be lost from the molecule to the medium and then re-enter the compound from the medium. This is a two-step intermolecular reaction. In the second possible mechanism, phosphate may not be lost but could be exchanged between the two sites in the same molecule, this being an intramolecular reaction.

3-phosphoglycerate

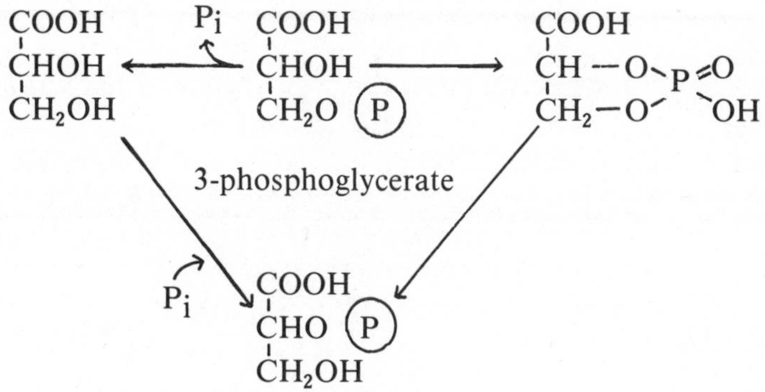

2-phosphoglycerate

If $^{32}PO_4{}^{3-}$ is added to the medium then in the inter-molecular reaction the $^{32}PO_4{}^{3-}$ would be taken up into the molecule and the 2-phosphoglycerate would be radioactive. In the intramolecular reaction this could not happen and in fact it is found that this is the case and the reaction must then be intramolecular. Once again this would be extremely diffi-cult to demonstrate without the use of isotopes.

Studies on Metabolic Pathways and Systems

By using radioactively labelled substrates or precursors a substantial amount of information can be obtained about the nature and rates of the chemical changes which occur during catabolic and anabolic processes. Again there are very many examples of such uses for isotopes and therefore only a few will be considered.

Using isotopes it is possible to determine whether a partic-ular compound, say A, can act as a precursor of another compound, say X, in the biosynthetic pathway forming X. Isotopically labelled A is administered to the system pro-ducing X and after a certain time interval, X is isolated, puri-fied and counted. If X is radioactive then A can act as a pre-cursor of it. However, if X is not radioactive this does not necessarily mean that A cannot act as a precursor of X because it is possible that the label may have been lost during the conversion of A to X or that A is failing to reach the enzymes synthesising X. Therefore, negative results obtained in this way must be treated with some caution.

An interesting example of this use of isotopes was the determination of the origin of the various atoms of the purine ring system. Many organisms can synthesise the purine ring system, found in adenine and therefore important in the structure of nucleic acids, from simple precursors. They can also break down purines and some organisms excrete uric acid as the degradation product.

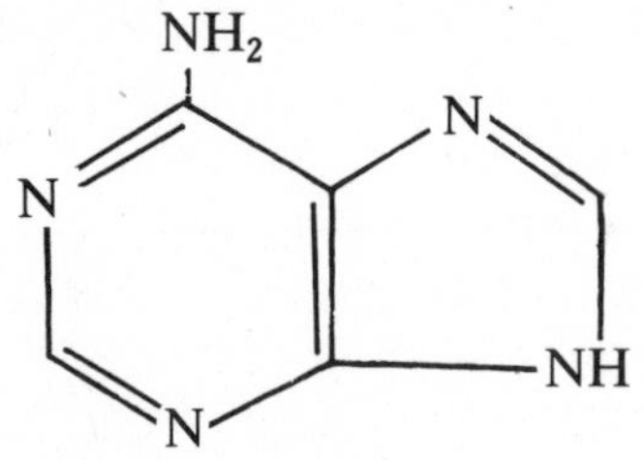

Adenine 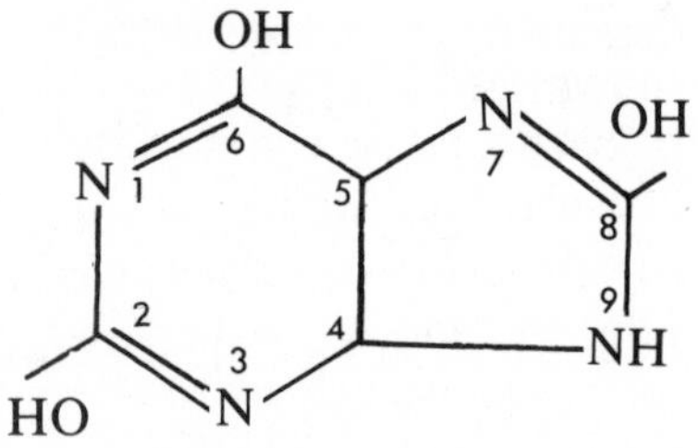   Uric acid

Uric acid can be extracted from the excreta, purified and chemically degraded into smaller fragments, these fragments coming from particular parts of the uric acid molecule and hence of the original purine ring system. This formed the basis for the method used to work out the origin of the atoms in purines. For instance, when radioactively labelled formate ($H^{14}COOH$) was fed to birds, they metabolised it along with the unlabelled formate normally available to them. Analysis of the degradation products from the uric acid showed that C-2 and C-8 were radioactive, and it could therefore be concluded that both these atoms are derived from the formate carbon atom. In a similar manner the origin of all the atoms was determined. The results of these investigations are shown below.

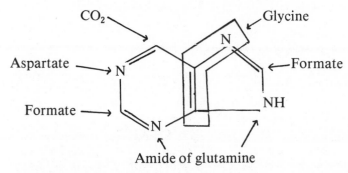

Amide of glutamine

Studies of the rates of reaction in metabolic pathways and identification of their intermediates are more difficult and depend upon certain specialised techniques such as a study of the variation of labelling patterns with time. This technique is based upon the fact that because metabolic pathways operate sequentially, an added isotopically labelled substrate will pass progressively along the pathway and there will be a time lag before all the intermediates are maximally labelled and this can be determined by counting. Thus if metabolism can be stopped at various time intervals from adding the label, the pattern of labelling will vary and it is apparent that the sequence of intermediates can perhaps be identified since those later in the pathway will take longer to become labelled. However, the amount of label present in any intermediate at a given time cannot be used reliably to indicate the sequence of intermediates because variation in the number of molecules in the 'pool' will also affect the intensity of labelling.

An additional value of these experiments is that the amount of activity in the compound through which the label enters the pathway declines with time, in contrast to the activity of the intermediates and end products. Thus, by using a variety of labelled materials and points of entry, the branches and convergences in the pathways can be identified.

It is important to remember that there are a number of problems with interpreting the results of these experiments. For instance, intermediates with only a transient existence will not be apparent since only very small quantities will be present. Again, the intermediate may spontaneously change to the next in the sequence or even to another, perhaps altogether non-physiological molecule, during sample preparation.

Perhaps the best known study of this type is the one carried out by Calvin and his associates in which they elucidated the sequence of reactions involved in the photosynthetic fixation of CO_2. A very much simpler version of their experiment is presented later in this book.

Isotopes can also be used to investigate the relative importance of alternative pathways. Glucose can be metabolised by at least three pathways and using isotopically labelled substrates the relative importance of these and the routes taken by the various carbon atoms can be determined. The three pathways concerned are:

(a) the Embden-Meyerhof-Parnas pathway (usually called glycolysis) which gives two molecules of pyruvate per molecule of glucose oxidised;

(b) the pentose phosphate pathway which yields three molecules of CO_2 and one molecule of pyruvate per molecule of glucose metabolised;

(c) the Entner-Doudoroff pathway which gives rise to two molecules of pyruvate per molecule of glucose metabolised.

Using specifically labelled glucose as substrate and degrading the products chemically for the location of radioactivity it has been found that the carbon atoms in the products are derived from specific parts of the glucose molecule and this is characteristic of each pathway. For example see p. 55.

Therefore, by using glucose labelled with ^{14}C at, for example, position C-1, and identifying, degrading and analysing the products (CO_2 and pyruvate) it is possible to determine the relative importance of each pathway.

Using such techniques it has been found that the pentose phosphate pathway is important during the biosynthesis of lipids and in the leaves of plants, while the Entner-Doudoroff pathway occurs principally in certain microorganisms.

Isotopes have also been used to study the distribution of

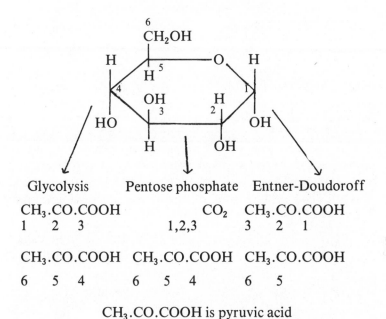

CH$_3$.CO.COOH is pyruvic acid

metabolic processes within cells and in this pursuit autoradiography has played a significant role. Radioactive compounds can be fed to cells and the site of their accumulation
can be detected by microautoradiography of light microscope or electron microscope sections. The photographs
showing the accumulation of ^{3}H-thymidine into cellular
DNA within the chromosomes of dividing cells are perhaps
well known. Similarly, the studies by Northcote using this
technique in order to show the synthesis and accumulation of
plant cell wall material in Golgi bodies and its progressive
movement to the cell periphery is another classic example.

Physiological Studies with Whole Organisms

The physiology of an organism comprises an interplay of a
large number of biochemical processes. A major difficulty in
attempting to study the physiological processes of organisms
is that the very fine balance in which they exist makes the
introduction of non-physiological events very likely as a

result of the experiment and this fact has not only seriously hampered physiological investigations but has led to the production of much erroneous information.

All organisms, whether animal or plant, from multicellular types down to the single cell, take up materials from their environment, distribute these materials throughout their structures, and metabolise them. In addition, they produce waste materials and useful products which are also transported within the organism and perhaps even exported to the environment. A knowledge of these transport phenomena is obviously essential to an understanding of physiology and indeed attempts at investigating them began over one hundred years ago. However, studies using non-physiological compounds chemically modified by, for example, halogenation or addition of dinitrophenol to make them distinctive is of limited use because these materials may not be metabolised in the normal way. The advantages of radioisotopes discussed in the introduction to this chapter are particularly significant regarding physiological studies and it will be constructive to consider a few examples.

In mammals, materials present in the food are absorbed from the gastro-intestinal tract into the blood stream and are thus distributed to other parts of the body. If a radioactive compound is added to the food it will be absorbed and distributed in a similar way to its non-radioactive counterpart, but it can be followed much more easily. By removing blood samples at various time intervals after feeding the radioactive materials and counting these samples, the rates of absorption from the intestine and removal from the blood stream can be found. For instance, amino acids can be shown to be taken up very quickly from the intestine, reach a peak concentration in the blood stream, and then fall in concentration as they are removed from the blood by cellular absorption. The main region of absorption can be shown to be the small intestine and studies of the uptake mechanism can be made. The effect of other materials present in the food, the effect of various environmental conditions and any departure from normal due to disease or other causes can be studied.

Sacrificing the organism used in such studies, removing

and extracting the organs and counting the extracts provides a very sensitive method for determining the distribution of the compounds under study. This may also be achieved by freezing the organism in liquid nitrogen, sectioning it with a fine band saw and exposing the frozen sections to X-ray film. Radioactivity in the various organs and tissues will blacken the emulsion as in a normal autoradiogram. For example, the accumulation of radioactive iodine in the thyroid gland can be shown in this way even in very large animals such as dogs.

Such studies are also possible with plants. Plants take up many elements from the soil in trace amounts and the importance of this can be demonstrated by growing plants under controlled conditions on media devoid of the element. However, much more information can be gained by allowing the plant to grow under more natural conditions on a medium containing radioactive trace elements. In this way not only can the material easily be shown to be absorbed, but its rate of absorption can be measured and the conditions affecting the rate of uptake and distribution of the trace element throughout the plant structure can be studied.

In higher plants, materials absorbed by the roots are transported to other parts of the plant largely via the xylem and since the rate of xylem flow is influenced by the rate of transpiration the use of isotopes in this way forms a very sensitive method for studying the conditions affecting transpiration.

Similarly, materials produced in the leaves as a result of photosynthesis are translocated down the phloem to the roots and other parts of the plant. Thus, if radioactive materials are applied or the plant is allowed to photosynthesise in $^{14}CO_2$, the rate and process of phloem transport can be studied by following the movement of the tracer.

This problem illustrates well the difficulty of studying physiological processes due to the necessity to interfere with the system. In this case, trouble is caused by the fact that the phloem system is under positive pressure and most sampling experiments necessitate interference with the system and thus destroy this normal state. This difficulty has been overcome by allowing aphids to feed on the experimental plant and if their heads are cut off, the remaining feeding tube (stylet) exudes phloem cytoplasm and can be used to continuously collect samples.

TURN-OVER RATES

The quantity of any tissue constituent may be reasonably constant in the adult organism under stable conditions, but this constancy is likely to be the result of a balance in the rates of synthesis and degradation of the constituent. The biological half-life of many materials is in fact quite short, a few seconds for ATP and only a few minutes for m-RNA. Even for cellular proteins it may only be of the order of 1-3 weeks.

Using isotopes two methods of study are available. First, an isotopically labelled precursor of the product under study can be fed to the organism and the rate of accumulation of the label in the tissue determined. Secondly, the store of a particular compound in an organism may be labelled by preliminary administration of a radioactive material; the subsequent loss of label from the organism on removal of the source of the label is then monitored.

One of the experiments described later in this book is concerned with the uptake and incorporation of $^{32}PO_4^{3-}$ by young mice and the results readily indicate that the phosphate slowly accumulates in bones over a period of about two weeks and then slowly falls over the next month or so. These slow rates are due to the particularly slow rates of turn-over of bony material.

CLINICAL APPLICATIONS

While it is accepted that ionising radiations are harmful, providing due care is taken it is possible to use both radiations and isotopes in the diagnosis and treatment of patients. The basic assumption is that the potential damage to the individual is outweighed by the benefits of diagnosis and/or treatment. While this important area will not be discussed in detail, a few of the more interesting applications will be mentioned.

Isotopes have been used to determine the size of various

body compartments such as the exchangeable sodium space in a tissue, the whole body water and blood volume, and the volume of localised pools of blood such as that present in the placenta.

The partitioning of material such as the amount of free thyroid hormone and the amount attached to serum protein can be determined and the body distribution of various isotopes can be assessed by whole body γ-ray scanners such as the γ-camera. By using materials specifically taken from the circulation by one organ system, the rate of functioning of that system can be investigated in an attempt to find malfunctions. For example, small colloids containing gold (^{198}Au) can be used in liver studies and heat-damaged red blood cells containing chromium (^{51}Cr) in spleen studies.

Studies on the sites and rates of accumulation of drugs, the rate of their metabolism and excretion are also routinely carried out using isotopically labelled drugs.

Ecological Studies

On a number of occasions isotopes have been used in various ecological studies and their use in determining insect population numbers by isotope dilution has already been mentioned.

Radioactive isotopes have also been used to follow insect movement. Trapped or reared insects are labelled, usually by feeding with or dipping into, radioactive material. They are released in large numbers and collected in traps arranged at a suitable distance and in a suitable pattern from the point of release. Continuous monitoring of the movement of non-flying insects such as flour beetles, and wood-boring or soil-inhabiting insects can be made if γ-emitting isotopes such as ^{60}Co are used.

Experiments along these lines are fraught with problems and not easy to interpret, but they have shown, for example, that house flies may move up to four or five miles for food and that in the transport of disease from sewers and cesspits to houses, cockroaches are not implicated but the biologists' friend *Drosophila melanogaster* is. Bear in mind that these studies are not limited to invertebrates; in fact the movement of birds and small mammals has been followed in the same way.

A different use for isotopes is in the identification of food chains and webs since the passage of an isotope from a labelled starting organism, through the population can in theory be followed. However, because of the considerable losses of activity at each stage in a food web and the fact that such a web has a complex branched nature, there is such a rapid dilution of activity that the starting organism has to be very heavily labelled indeed. Nevertheless, some useful results have been obtained and it has been demonstrated, for example, that drone bees enclosed in wire cages with a solution of labelled sucrose do not become radioactive since they are fed by free-flying workers. Similarly, certain ant colonies may become radioactive because of their habit of 'farming' aphids which have fed on labelled plants.

Other Uses

It is not surprising that a tool of such great value as that provided by isotopes has been used in a very wide variety of investigations and it may be useful to mention he a few of the less well known or less important of such applications.

The fact that radioactive isotopes are present in all organisms and that the total amount of these begins to fall by normal decay upon the death of the organism has been used to estimate the age of fossilised or otherwise preserved material. The general principle of radioactive dating is that one very carefully determines the amount of radioactive isotope left in the material knowing the amount originally present and the half-life of the isotope. The age of the sample can then be calculated. Generally speaking, ^{14}C is the isotope used for short-term dating, and ^{40}K or the uranium isotopes for longer term work.

Unfortunately, the method suffers from some potentially serious errors particularly the difficulty of accurately determining the half-life of the isotope involved. The assumption that the level of isotope present in the environment (and hence in living things) has remained constant throughout geological time is also being questioned at the present time. Dating with tree rings (dendrochronology) indicates that this assumption is not valid and corrections need to be applied to the dates obtained.

The process of nuclear activation has been used to identify certain elements, particularly minor components such as arsenic and manganese in animal foods. If the material in question is bombarded by nuclear particles from a deuteron accelerator, radioactive products are produced which can be identified by their characteristic decay radiation.

There are also of course, a large number of procedures in which emitted radiation, rather than a radioactive isotope, is important and perhaps the clinical use of X-radiations is best known. While this subject is really outside the scope of this book, two other interesting uses might be noted. γ-radiation from ^{60}Co and ^{137}Ce has been used to sterilise plastic medical ware and even food and drugs, although in this case one usually has to be content with partial sterilisation aimed at reducing the decomposition rate since high radiation levels may adversely affect the material.

A related use to which radiations have been put is in the attempted production of mutations, particularly of micro-organisms for biological research. Attempts to produce new strains of commercially important plants by this method have often failed, the barley variety 'Proctor' being one of the successful examples.

The discussion in this chapter has been purposely restricted to the use of radioactive isotopes because these have been perhaps the most useful to date in biological experimentation and it is the area with which the experiments in this book are concerned.

There are, however, numerous examples of the use of non-radioactive nuclides other than the naturally predominant ones, particularly in biochemical experiments. The fact that these isotopes are not radioactive means they cannot be counted by Geiger-Muller tubes or by scintillation counters. They can be determined by virtue of their heavier mass using a mass spectrometer but these instruments are extremely expensive, not easy to operate and require the compound to be pure. They are, however, very sensitive and enable one to use these nuclides as tracers in the same way as radioactive nuclides.

Deuterium ('heavy hydrogen') is a good example of one of these and it has been used for instance, to study the specificity of enzymes with regard to the two hydrogen atoms of the nicotinamide residue of $NADH^+$.

There are no readily available radioactive isotopes of oxygen or nitrogen which have half-lives of useful length and here again the heavy stable isotopes ^{15}N and ^{18}O must be used. ^{15}N has for example, been used in studying inorganic nitrogen metabolism, and ^{18}O in studies of the metabolism of oxygen containing groups, particularly the origin of hydroxyl groups and the oxygen evolved in photosynthesis. The value of radioactive isotopes in studies of metabolism may perhaps be gauged from the fact that our knowledge of inorganic nitrogen and oxygen metabolism is significantly less than in those areas for which radioactive isotopes are available.

CHAPTER 5

THE BIOLOGICAL EFFECTS OF RADIATION

The biological effects of ionising radiations are quite complex and not well understood. The visible or demonstrable symptoms resulting from these effects occur at a variety of levels of biological organisation and after various lengths of time, although, of course, they are all derived from the chemical changes occurring at the time of impact of the radiation. These processes are at present under quite extensive study because of the necessity of treating people accidentally injured by radiations, and for military reasons; but also because of the basic information about genetics and cellular repair mechanisms which might be obtained.

Radiation absorbed by aqueous systems such as cells can result in molecular excitation, ionisation, or even the production of free radicals. It is the latter which present the greatest biological danger because of their tendency to induce oxidations and reductions and in particular their ability to initiate chains of such reactions. For the purpose of this discussion it is convenient to divide the effects of radiation into those appearing at the biochemical, cellular and whole body levels. The most important of these changes are shown in Figure 17.

Biochemical Effects

It seems to be the case that biochemicals can be affected either directly or indirectly by radiation. Direct action results

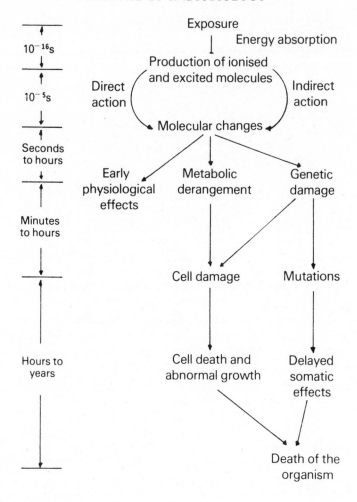

Figure 17. Radiation effects and their sequence

from the effects of radiation on the biochemical itself. The importance of the damage caused depends upon the precise change produced in the molecule and the ease with which it can be replaced or repaired. Damage to molecules present in large numbers, glucose for example, is not usually important unless the product happens to be poisonous. Even the destruction of an enzyme molecule is not very important since these can usually be replaced by the normal mechanism of protein synthesis. Damage to DNA is generally considered to be the most important direct effect of radiation since this is the fundamental store of information for the operation of the cell and a change in it is likely to have serious consequences. However, it is well to bear in mind that in any given cell the majority of the DNA present is never used and changes in this component might go unnoticed unless they affected the process of DNA replication. In addition, it is now apparent that the cell can carry out repairs to its DNA in order to counteract changes arising in all manner of ways.

Indirect effects result from the attack, by radicals produced in the aqueous environment, on important biochemicals. Perhaps the majority of these attacks are by radicals produced from water and the main routes by which they are derived are given below.

$$H_2O \quad H_2O^- \rightarrow \boxed{H^\cdot} \;+\; OH^-$$

$$H_2O \rightarrow H_2O^+ + e^- \qquad \qquad \downarrow$$

$$H_2O$$

$$\rightarrow \boxed{OH^\cdot} + H^+$$

$$H_2O \rightarrow H^\cdot \;+\; OH^\cdot$$

These indirect processes are important in cells because of the aqueous nature of the cell and because large molecules like proteins and nucleic acids and structures such as membranes are susceptible to them. However, they are not as serious as one might expect since there is evidence that biologically important molecules are partially protected *in vivo* and in addition many radicals produced in the cell are lost in trivial reactions.

It should be apparent from the above that quite a large number of radiation impacts might occur and do no serious biological damage, yet, on the other hand, the first impact of even a small dose of radiation might result in changes to the cell and indeed the whole organism. Those changes which initiate events ultimately leading to serious biological damage are sometimes called primary lesions.

There are certain important and well-studied biochemical changes which are suitable for discussion here since they usefully illustrate the sort of effects radiation can have on molecules.

Proteins are macromolecules of considerable importance and their size and number ensure that they will frequently be the sites of impact of radiation. Cleavage of bonds has often been noted and it seems that both the peptide bond and the other bonds (principally the so-called sulphur bridges) involved in holding together the coiling and arrangement of the amino acid chain into the shape characteristic of each protein, can be broken. In addition, changes in the amino acid side groups of the protein can result in cross linkages developing both within a single molecule and between adjacent molecules. Bearing in mind that for most functions of proteins a particular shape of each molecule is required one can easily see why quite low doses of ionising radiation cause losses of protein function.

A wide range of effects has been seen as a result of irradiation of DNA suspensions or of living cells. Fortunately, many of the more serious of these occur only at high dose levels. Alterations to individual bases due to direct effects of radiation seem fairly frequent and these have serious consequences regarding base pairing and DNA function generally. Radiation impacts also seem to be able to cleave the pentose-phosphate 'backbone' of the molecule quite easily although, because of the ability of the cell to repair this damage and the unlikelihood of two breaks occurring close together, actual cleavage of the DNA helix is rare. Separation of the two strands of the helix is also unlikely because of the very large number of hydrogen bonds which are present and which would presumably need to be broken.

Lipids are another group of compounds which are easily affected by radiation. Lipid molecules are very sensitive to oxidation, particularly if they are unsaturated, and to cleavage of their hydrocarbon chains. These changes can be of great significance since lipids are a major component of cell membranes, and are principally involved in the prevention of free diffusion of materials through membranes. As we shall see in the next section many radiation induced cellular and physiological changes can be traced to failures in the functioning of membranes.

Cellular Effects

Cellular effects of radiation can be broadly divided into two types, effects on cell division and effects on other aspects of normal cellular metabolism. This separation is largely a matter of convenience and some overlap and inter-dependence occurs.

As a result of the changes to DNA and proteins described previously, visible damage to chromosomes can sometimes be seen. Such damage can involve a physical cleavage of a chromosome with the fragment not containing the centromere either remaining free or becoming attached to another chromosome. It is also possible for the fragment to become re-attached to the original chromosome either in the original orientation or after inversion. Exchange of parts of different chromosomes is much less frequent since it requires the cleavage of two chromosomes and the exchange of the broken parts. All of these exchanges are potentially serious since they affect either the number of genes in the chromosomes or their sequence.

The cell division process is also affected not only because some of the changes discussed above affect the ability of the chromosomes to duplicate and separate during cell division, but also because the mechanism of the cell division process is adversely affected in largely unknown ways. As one might expect the majority of cells are most prone to radiation damage during that stage of interphase when the DNA is duplicated and during the process of cell division itself.

Low levels of radiation lead to a delay in cell division from which the cells gradually recover and divide as normally. The delay is probably due to a temporary halt or slowing down in the process of chromosome assembly, or to an increase in stickyness which makes chromosome separation more difficult.

Higher doses of radiation may inhibit cell division permanently, the cells not recovering, eventually dying and being replaced. In some cases, the cells divide normally for as many as ten divisions but then die and before dying may grow into giants up to thirty times their normal size. This is taken to indicate that their general metabolism is functioning normally but that their division mechanism probably is not.

If chromosome breakage has occurred then any fragments produced without a centromere will fail to assort properly at cell division and will become randomly distributed among the daughter cells. After a few division cycles the cells are usually grossly enlarged with abnormal complements of whole chromosomes and fragments. These abnormalities usually lead to an early death of the cells.

Instant death requires very high irradiation and the cell very clearly suffers extensive damage. Nuclei and other organelles swell and distort and the DNA can be seen to form droplets and coagulate.

For reasons discussed in the previous section, effects on general metabolism of cells are not likely to be due to changes in cellular enzymes. Rather they seem to be due to changes in cellular membranes and the previously supposed sensitivity of respiratory enzymes to radiation is more likely to be an effect on the mitochondrial membranes to which these enzymes are attached. Radiation damage to membranes is potentially very important because it seems that after a certain dose level membrane degradation becomes autocatalytic probably because of the formation of chain reactions among lipid radicals within the membrane.

A long list of these membrane based changes has been accumulated and a few examples will serve to illustrate the sort of effects and their consequences. Nerve cells are particularly prone to membrane damage because of the cell's dependence upon the membrane for electrical insulation. One finds as a result of irradiation, irregular uncontrolled nerve impulses of abnormal intensity. Control of active transport through the outer membranes of the cell becomes difficult and one finds that, for example, red blood cells fail to control their electrolyte balance and plant roots have difficulty in absorbing ions. Intracellular membranes are also affected but fortunately, bearing in mind the consequences of lysosome breakage, only at fairly high doses.

Whole Organism Effects

The study of the effects of radiation on whole organisms is difficult for a variety of reasons. The effects are very diverse in nature and may influence any aspect of an organism's physiology or behaviour. It is unfortunate that they vary qualitatively and in particular, quantitatively, in different species and they can be very delayed. Even now supposedly unharmed Japanese are developing conditions which are almost certainly due to radiation exposure in 1945 and indeed some of the effects of radiation may only become apparent in generations subsequent to the one exposed. Another problem is that the age of the irradiated organism and sometimes the conditions under which it is grown also affects the result obtained after irradiation. The presence of these variable factors makes attempts at investigating the mechanisms of whole body radiation effects, by comparative experiments, very difficult.

At one time radiation effects were separated into two groups, those changes seen immediately after irradiation (sometimes called the acute radiation syndrome) and longer term conditions. It was considered that below a certain threshold of radiation dose only the shorter term, probably repairable, effects prevailed. It is now, however, accepted that there is no threshold for long-term effects and thus very low doses may have the most serious long-term consequences. This fact is of considerable importance to radiation workers and has led to a downward revision of maximum permissible doses of radiation. The results of exposing mice to radiations of different doses are shown in Appendix 3. We will briefly discuss these in turn.

High doses of radiation usually kill the organism within a few days and, in the case of mammals, the most probable cause of death varies with the dose absorbed. Very high doses (15 krad) lead to widespread death of cells through metabolic malfunction, lower doses (10 krad) induce lung or nervous failure and doses of about 1 krad cause death because of changes in the gastro-intestinal system which result in uncontrolled loss of Na^+ and water.

Radiation sickness is the general name for the condition

which occurs after doses of a few hundred rads. It results initially in headaches, vomiting and general debility and leads on to intestinal upsets and gross hormonal changes. These problems seem to be largely due to effects on membranes, of which those effects on cells of the central nervous system, gastro-intestinal tract and endocrine systems seem most important. The elevation of circulating hormone levels due to leakage from cells of the endocrine glands is so characteristic that it has been used as a semi-quantitative indicator of the intensity of exposure.

There is sometimes a false recovery phase of up to twenty-four hours during which the organism appears to recover. This is followed by the second stage of the condition, a period of variable length which is characterised by haemorrhages, anaemia, infection and malnutrition. All of these effects can be traced to degeneration or death of various generative cell systems possibly because their high rate of cell division makes them particularly sensitive to radiation damage.

The failure of new cell production for the blood system has serious consequences, lack of oxygen causes debility and a low level of leucocytes readily allows infection. Fragility of capillary walls and loss of platelets and clotting factors means that frequent haemorrhages occur. Malnutrition develops because of a failure of the body to replace the food absorbing cells of the intestine after they have been sloughed off during the normal process of peristaltic movement.

Death can occur at any time during radiation sickness, precisely when (and if) it occurs depending upon the exposure, the species concerned and the age and general health of the exposed individual.

Irradiation is very unusual amongst the causative agents of disease in being able to produce effects a considerable time after exposure. Most of these delayed effects are genetic in origin but some may be due to long-term somatic changes such as the slow build up of toxic materials. Perhaps the most important of these changes are malignancies in various forms including skin and eye damage, leukaemia and thyroid cancer. One also finds a statistically proven, but largely unexplained, decrease in longevity and growth and a variety of deleterious effects on the offspring of the irradiated individual. It is particularly unfortunate that the generative cells of the mammalian gametes are very sensitive to radiation and in the case of females, once these cells are killed, they cannot be replaced.

Protection and Repair

It has been mentioned previously that all radiation impacts do not produce permanent damage to cells because radiation can be absorbed in non-harmful ways, because the cell seems to provide a certain amount of protection and because in some cases radiation-induced damage can be repaired.

Experiments frequently demonstrate that a particular chemical or structure requires greater exposure for a particular effect when it is present *in vivo* than when *in vitro* and this has led to the idea that protective mechanisms exist within the cell. A number of chemicals have been shown to act as radioprotectors probably by competing with the bio-logically important molecules for free radicals of aqueous origin. The most effective radioprotectors contain sulphur or more usually nitrogen and radioprotection is perhaps one of the roles of some of the mysterious compounds of unknown function present in the cytoplasm. It certainly seems likely that the protein associated with DNA has this role amongst others, and the relative radioresistance of many invertebrates has been attributed to their tendency to retain free amines in cell and tissue fluids in order to maintain the required osmotic potential. Vertebrates retain ions for the same purpose and may not gain this additional benefit of partial radioprotection.

One can imagine organisms faced with potentially dan-gerous background radiation throughout their evolution developing two kinds of repair mechanism, the neutralis-ation and repair of secondary effects resulting from the initial damage to cell molecules, and the replacement of dead cells. That these repair mechanisms exist was known a long time before much was understood about them, since it is easily demonstrable that a given dose of radiation split into two or more periods is less effective than if it is given in a single dose. The obvious conclusion is that cells are able to partially repair the damage caused during the period of non-exposure or 'rest' and hence deal with the total dose more easily.

Most of the mechanisms available to cope with radiation damage to biochemicals are not special repair mechanisms at all but are the normal replacement mechanisms evolved to maintain the structure and function of the cell despite a constant turnover of its components. Thus, three weeks from now half of the reader's liver protein will have been degraded and resynthesised anyway, and low levels of radiation may only necessitate a very slight increase in the rate of such processes.

DNA however, is a molecule that is repaired rather than replaced possibly because the chances of error in the latter are greater than in the former. The system involved is a complex one containing enzymes which can move along DNA helices monitoring them for chemical abnormalities, and other enzymes cutting out and replacing individual bases or sections of DNA strand as necessary. It now seems possible for diploid cells to repair a DNA helix in which both strands are affected at the same point by using the other helix as a template for the repair.

In large multicellular organisms an irradiated area should be surrounded by healthy tissue and it seems that recovery of the irradiated cells is stimulated by nutrients or growth factors from these healthy cells. If, however, the irradiated cells should die the volume occupied by them can be repopulated by division of the healthy cells, assuming that the dead cells were not too specialised (nerve cells, for example). In addition, it is now known that for many cell types, small numbers of cells are continually circulating in the blood and lymph and will act as 'seeds' when regions of the tissue have been destroyed. This process, also important in the development of cancers, is termed metastasis.

In conclusion it can be justifiably said that every radiation impact, even a single β-particle, may be of great importance and consequence if it hits a vital component at a particular time and in a particular manner. However, the danger involved is roughly proportional to the amount of radiation received over the lifetime of the individual and with care even individuals who use isotopes daily should not receive what has been calculated and shown by experience to be an excessive dose.

To balance these sombre notes it is useful to remember that evolution has been helped in no small way by the fact that even normal background radiation induces mutations and also to remember that malignancies can be treated as well as produced by radiation.

CHAPTER 6

THE SAFE USE OF RADIOACTIVE
ISOTOPES IN TEACHING
EXPERIMENTS

Apart from any dangers due to the chemical nature of the materials used or to the apparatus or techniques involved, experiments using radioactive isotopes present special problems of their own. These problems centre, of course, around the fact that unstable isotopes disintegrate producing potentially harmful radiations.

A difficulty peculiar to the use of radioactive isotopes is that as these emissions are not detectable by any of our senses; they are an insidious danger. Another problem is that prevention of the possible effects of irradiation is essential since cures are difficult or more usually impossible. Fortunately, cells have inbuilt repair mechanisms which have evolved to cope with the effects of natural radiation and other types of damage and which usually are able to cope with experimental levels. Only excessively high levels of activity or experimental accidents are likely to lead to serious damage.

The quantities of isotopes used in most biological experiments do not present a great hazard to health. It is perhaps the case that isotopes emitting β-radiation are the most frequently used and it is useful that the radiation of these isotopes is comparatively weak, being readily absorbed in liquids and solids. Thus the β-radiation of ^{32}P, while having a maximum range of about 6 m in air, has a range of only about 6 mm in water. Open solutions of these isotopes do not therefore emit as much radiation as one might expect and in closed containers the emission is severely cut down. A point always worth bearing in mind when using isotopes is that the intensity of the radiation declines in inverse proportion to the square of the distance from the source. As one approaches a source of radiation therefore, the potential danger from it rises sharply and contamination of the skin in particular is to be avoided if at all possible. From a human safety point of view, most β-radiation striking the body is absorbed in the first few millimetres of skin. However, it should be borne in mind that β-particles such as those produced by ^{32}P can cause X-rays of long range to be produced by glass or metals and this may constitute an entirely different health hazard.

Isotopes emitting only γ-radiation are not often used in biological experiments because of the difficulties in counting them and for other reasons. However, some of the commonly used β-emitting isotopes such as ^{24}Na, also produce γ-radiation and have to be used more carefully on this account. The health hazard in using these isotopes is much greater than for β-emitters because the radiation, even from a source outside the body, is capable of passing completely through it and more serious damage could therefore occur. α-emitting isotopes are rarely used in biological studies because they are largely confined to very heavy elements such as radium. These isotopes constitute little in the way of hazard when present outside the body because of the short distance that α-particles travel in air.

The situation in which all isotopes can be dangerous is when they become absorbed by the body following ingestion, respiration or by entry through cuts and abrasions. After absorption isotopes are in most cases fairly evenly distributed throughout the body and become intimately mixed with and lie next to molecules and structures which may be sensitive to radiation damage; for example nucleic acids, proteins and membranes. The radiation received by these materials is intense because of the close proximity of the disintegrating atoms, and even α-particles become dangerous because of the high degree of ionisation they produce in surrounding molecules.

The extent of the hazard from ingested isotopes depends upon a number of factors principally the chemical nature of the compound taken in and its distribution within the body. For example, tritium gas ($^3H_2$) is virtually harmless as it is not incorporated into biochemicals and is eliminated very quickly from the body. Tritiated water (^3H_2O) on the other hand, presents about 10^5 times the hazard arising from tritium gas since it is easily absorbed through the skin, readily incorporated into biochemicals and not purposefully excreted by the kidneys. ^3H-labelled organic compounds vary in their potential danger and in particular, bases such as thymine are very dangerous because of the likelihood of their incorporation into vital and very susceptible molecules such as nucleic acids present in, and associated with, chromosomes and ribosomes.

The rates of turn-over and excretion of the isotope are also important. While ^{14}C has a decay half-life of about 5570 years, its biological half-life (for most compounds) is in the region of 30 to 40 days; in other words after this period of time from the ingestion of the isotope about half of it will have been excreted. A great deal depends on exactly where the ingested isotope becomes deposited since molecules involved in the central nervous system, in bone and in DNA itself tend to be long lived and are not readily broken down and resynthesised and hence do not yield many opportunities for isotopes to be eliminated.

Uneven distribution of the isotope throughout the body can increase the hazard due to taking in even a small amount of radioactive material. A number of isotopes have this unfortunate tendency and must be treated with increased respect during experiments. Well-known examples are the accumulation of calcium and phosphate in bones and iodine in the thyroid glands. The maximum permissible quantities of these isotopes that can be used in single experiments are correspondingly lower than for other isotopes.

Another problem arising from ingested isotopes lies in the nature of their decay products; ^{32}P for instance decays to ^{32}S and not to the non-radioactive and more common ^{31}P. Thus if ^{32}P is ingested and some of it is used in the biosynthesis of cell components, there will be a continuous degradation of these to sulphur atoms. This will drastically alter the chemical properties and stability of these compounds and it could be serious if it occurred in nucleic acids or in molecules involved in structures such as cellular and organelle membranes.

For the reasons described above it is important that the external and internal contamination of the body by radioactive isotopes should be prevented as far as possible. The following precautions should be taken to ensure this.

1. No unnecessary materials such as books, papers, clothes or similar articles should be placed on the laboratory benches. These items should preferably be left outside the laboratory.

2. Laboratory coats should be worn at all times. These should preferably be ones kept solely for use in radioactive laboratories and which are regularly washed to prevent build up of contamination. The hands should be protected with rubber gloves and the eyes with safety glasses. Washing the gloves before removing them will prevent the build up of contamination on them and the possible transfer of this to the body or to apparatus in subsequent use.

3. Eating, drinking, smoking and the application of cosmetics is absolutely forbidden in the laboratory. Disposable paper handkerchiefs should be used if these are needed.

4. All wounds below the wrist, even if covered by plasters or bandages, should be reported to the class supervisor who will decide if it is safe to handle isotopic material. Wounds provide points through which isotopes can easily enter the body.

5. Mouth operations of all kinds are strictly forbidden in the laboratory. Pipette bulbs, or other kinds of automatic dispensing devices, should be used. Adhesive labels or other materials should not be inserted in the mouth.

6. All operations involving high energy β-emitting or γ-emitting isotopes should be carried out behind suitable screens. Fume cupboards are useful for this.

7. If high energy emitting isotopes are used, or if isotopes are used over long periods, a film badge should be worn in a prominent place and developed at intervals of not less than one month and preferably fortnightly.

8. Routine analysis of the urine, by scintillation counting, is desirable if millicurie quantities of ^{14}C or ^{3}H are being used.

9. Dispensation of isotopes of high specific activity, as for example from stock solutions, should be made by the class supervisor to reduce the possibility of spillage of extremely active solutions.

10. If active solutions are spilled on the skin, it should be well washed with soapy water and the accident reported to the class supervisor.

11. On leaving the laboratory and after removing the coat and gloves, the hands should be washed and the clothes monitered for contamination using a suitable hand monitor.

It is essential to avoid, as far as possible, contaminating the environment, particularly that of the laboratory. Careless work could, over a period of time, result in the accumulation of an isotope to a potentially dangerous level and also produce a high background count rate or contamination of glassware; both of these problems leading to a reduction in the usefulness of experimental results. The following precautions should be observed to overcome this possibility.

12. All work should be carried out on or over surfaces which are easily cleaned e.g. plastic laminates, trays or even glass. These should be covered with absorbent paper which is treated as contaminated waste at the end of the experiment.

13. Work involving radioactive gases or powders should be carried out in fume cupboards or in special apparatus. Any such apparatus should have tight fitting, greased joints.

14. Any spillage occurring during the experiment should be mopped up with absorbent paper, the area sluiced with water or solvent and wiped again. The paper is then treated as contaminated waste. Rubber gloves should obviously be worn during this operation and the class supervisor informed of the accident so that he can monitor the area for contamination before allowing work to proceed.

15. To avoid spitting of solutions being evaporated to dryness the heat intensity should be reduced when the sample is nearly dry.

16. Whenever radioactive materials or solutions are carried they should be, if at all possible, in a tightly closed container carried within or on another container to minimise the possibility of spillage and/or breakage.

17. It is imperative that contamination of the counting room is avoided. Only correctly prepared samples should be taken into the counting room and rubber gloves should be removed before entering the room.

Correctly prepared samples are those which are either dissolved in scintillator fluid and kept in screw-cap bottles or sealed tubes, dried down on to planchets, or, if loose and solid, placed on to planchets and covered with Sellotape.

18. At the conclusion of the experiment all radioactive liquids should be poured into clearly labelled bottles kept for the purpose, and never poured directly down the sink. Radioactive solid waste should be placed in marked pedal bins or polythene bags. All these materials should ultimately be disposed of according to the authorised procedures.

19. Radioactive glassware and apparatus should be well washed and kept separate from non-radioactive materials. If there is a likelihood of active materials adhering to the apparatus it should be well washed in carrier solutions, then if necessary in chromic acid or detergent solution. Planchets and plastic scintillator bottles are treated as disposable and should not be washed.

20. Radioactive samples that are to be stored for further use should be placed in tightly stoppered containers labelled with yellow isotope warning tape. They should be labelled to indicate the type(s) of isotope involved, the approximate activity, the date of placing in store and the owner's name. They should be placed carefully so that they are not knocked over and broken.

21. The area over which the experiment was performed should be monitored to check for contamination as a final act before leaving the laboratory.

The fact that radioactive materials are potentially dangerous to use has led to the production of a plethora of Government regulations concerning their purchase, transport, storage, use and disposal. Perhaps the most important of these documents is the 'Administrative Memorandum 2/76' together with its guidance notes issued by the Department of Education and Science, and the 'Code of Practice for the Protection of Persons Exposed to Ionising Radiations in Research and Teaching' issued by the then Ministry of Labour in 1974.

It is not really necessary for this book to list these and other regulations and recommendations, and anyone interested in using isotopes in teaching experiments is advised to obtain copies of them and any additional advice he needs from the Department of Education and Science. The excellent book by Dance (1973) listed in the bibliography provides a useful summary of these regulations.

CHAPTER 7

EXPERIMENTAL TECHNIQUES

This chapter discusses four techniques which occur quite frequently in the experimental section of the book, the discussion is included here to avoid undue repetition.

Preparation of Planchets for Counting in an End-window Geiger-Muller System

To obtain the best results using the Geiger-Muller tube and planchet technique, experience and good manipulative skill are required—two characteristics not shown by many students. If badly prepared, the geometry (i.e. shape and distribution of the sample) of different planchets may vary considerably and this affects the count rate and reduces the accuracy of the results, particularly where emitters of weak β-radiation are used. Since the efficiency of counting is low anyway, the interpretation of all the results obtained by this method of counting must be done carefully.

For reproducible results two points should be borne in mind:

(a) approximately equal weights of sample should be plated (i.e. dried on to the various planchets); this being most important if particulate or dense samples of different composition are being counted.

(b) equal volumes of solution should be used and if possible plated all at once in order to form a uniform film. It is also important that the whole of the planchet is covered by the sample film.

In practice, the first condition is difficult to satisfy since in many experiments the weight of material to plate cannot be found easily. Normal size planchets are 2·5 cm in diameter and will hold about 1 ml of liquid. If more solution than this has to be plated it can be evaporated to a smaller volume before plating although this may be difficult with aqueous solutions, especially on a class basis, and it may lead to errors due to adsorption on container surfaces. Alternatively, the solution may be evaporated on the planchet using repeated small additions. This has the disadvantage that much of the material will accumulate round the edges of the planchet and the risk of materials spitting out of the planchet or 'climbing' over the edge will be increased. These difficulties must be balanced against the relative simplicity of the latter procedure and in any case really accurate results are not very often required in student experiments.

Small volumes of aqueous solutions do not 'wet' planchets well but tend to accumulate in small drops; this does not lead to the production of an even thin film. In this case, as with plating aqueous solutions in general, better results are obtained if one drop of dilute detergent is added before evaporation. Aqueous solutions can also be made up in 1% detergent instead of water if required.

The evaporation of aqueous solutions on to planchets may conveniently be carried out on a hot plate or by using an infra-red heat lamp. In either case, care should be taken to avoid getting the liquid so hot that it foams and spits. When the sample is nearly dry the heat should be reduced to avoid the material 'jumping' out of the planchet. This is particularly important with particulate materials such as yeast cells or red blood cells.

If loose material, for example leaf discs or powders, are to be counted the planchet should be covered with adhesive tape in order to prevent spillage and contamination of the counting instruments.

Note that used planchets are always treated as disposable.

Preparation of Samples for Scintillation Counting

If only a small number of samples or experiments are envisaged it is probably best to buy ready-made scintillator from one of the radiochemical suppliers. These are not very expensive and have the advantage of constancy of composition and properties. The scintillation mixture NE.240 from Nuclear Enterprises Ltd has been used by the authors for some of the experiments listed later.

It is of course possible to take more trouble over scintillant preparation and the bibliography contains several references discussing this subject in more detail than we can here.

One can divide sample preparation systems into homogeneous and heterogeneous systems of which the former occur more frequently. Perhaps the most often used mixture (or 'cocktail') contains

6 g PPO (2,5-diphenyloxazole)
0·3 g POPOP (1,4-bis-2-(5-phenyloxazolyl)-benzene)
100 g Naphthalene

dissolved in one litre of toluene.

Toluene is a very popular solvent because of its efficient energy transfer, high purity and low cost. It does not, however, readily accept aqueous samples and mixtures are sometimes modified by the addition of materials designed to improve this. If 300 ml of peroxide-free 2-ethoxyethanol is added to the above mixture, incorporation of aqueous solutions is greatly improved, although some proteins and buffers may precipitate out of it.

Mixtures based on dioxan (1,4-diethylene dioxide) are frequently used for aqueous samples despite a potential decrease in counting efficiency. The one listed below for example, is capable of accepting up to 20% water but may become cloudy at certain salt concentrations and is sensitive to environmental temperature.

6 g PPO
0·3 g POPOP
100 g Naphthalene

dissolved in one litre of dioxan.

Bray's solution contains:

4 g   PPO
0·2 g   POPOP
100 ml Methanol
20 ml Ethylene glycol

made up to one litre with dioxan. It is popular because of its high solubility for aqueous solutions and high counting efficiency. However, it may also react adversely to certain salt concentrations.

It is worth noting that there have been a number of developments in scintillator chemicals (fluors) in recent years in an attempt to improve energy transfer to the modern photomultiplier tubes. PPO (2,5-diphenyloxazole) and the wavelength shifter POPOP 1,4-bis-2-(5-phenyloxazolyl)-benzene can be usefully replaced by single chemicals such as MSB (a styrene/benzene derivative) or PBD (another phenyl-oxazole compound). At present perhaps their relatively high cost prohibits their widespread acceptance.

In heterogeneous sample systems the cocktail and sample do not present a single phase. These systems lack the intimacy of contact, stability, reproducibility and high efficiency of homogeneous systems but may be necessary where particulate materials or low specific activity material is to be counted. It is also more difficult to determine their counting efficiencies.

Particulate material cannot be reproducibly counted simply by placing in a cocktail in the normal manner because of its tendency to sediment. Suspensions of the material can be easily made by using 3–5% Cab-o-Sil (trademark of the Cabot Corp.). This is a thixotropic silica material which can support 1 g of the material to be counted ($BaCO_3$, bone ash, protein precipitate, etc.) per 10 ml cocktail.

Because of the problems mentioned above, attempts are frequently made to solubilise particulate material before counting. Several very good solubilisers are commercially available and the in-vial combustion technique described below is also satisfactory.

After placing the sample (0·2 ml or 20 mg) into a vial, 0·2 ml of 70% perchloric acid and 0·4 ml 100 volume hydrogen peroxide are added. The vial is then heated, with a loosely screwed cap, at no more than 80°C for 20-30 minutes. A scintillation cocktail not containing POPOP may be added to the digested material and the sample counted in the normal way.

A procedure which is becoming increasingly popular is emulsion counting in which undissolved samples are counted as a detergent-based emulsion rather than as a suspension. The mixture

3·2 g PPO

0·08 g POPOP

dissolved in 800 ml toluene and 400 ml Triton X-100, is frequently used since samples are counted at high efficiency, the system is easy to prepare and it is very resistant to quenching. One has to be careful to check, however, that all sample bottles show the same emulsion state.

As a final point it is perhaps advisable to note that most cocktails contain materials that are inflammable, volatile, carcinogenic or otherwise harmful. Careful storage is therefore required and dispensation by automatic means is advised. There is in fact no need for cocktail dispensation to be very accurate as this seems to have little effect on the count rate except at low volumes.

The process of sample preparation for liquid scintillation counting is usually straightforward. The sample, most often 1 ml or less, is pipetted into the bottle or tube, dried or solubilised if necessary, and a suitable volume, normally up to 20 ml, of scintillator is added. The bottle cap is then screwed on as tightly as possible before mixing the bottle contents. It is essential to fit the cap tightly because the low surface tension of most scintillators allows them to escape relatively easily past the bottle cap. The sample is mixed well and if necessary cooled for a minimum time of 1 hour in order to reduce chemiluminescance, before counting.

Materials of known activity for use as radioactive standards for scintillation counting can be either prepared by the user or purchased ready made. In the former case the radioactive material is accurately weighed or dispensed into a scintillation bottle, dissolved in scintillator and counted in the normal way.

Users of scintillation counters are always examining ways of reducing the costs incurred during the counting of large numbers of samples. For example, many people now use smaller volumes (3–5 ml) of scintillation cocktail and accept the slightly lower count rates. Polythene insert tubes are frequently used to hold this lower volume in the centre of a scintillation vial. Full-size polythene vials are now sometimes used in place of glass ones with a reasonable saving in cost and with the advantage of low background count rates. Contrary to the manufacturers' instructions, they can be re-used if a check is first made for contamination due to absorption, and for swelling due to interaction with the solvent; it is doubtful, however, if the saving in cost merits the effort involved.

Another variation is to absorb the sample to be counted in a small glass fibre disc and to place this in the base of the vial covered by 2–3 ml of cocktail. It is claimed by some workers that, if the radioactive material is aqueous and a toluene cocktail is used, then this can be decanted off and used in other experiments.

Autoradiography

The preparation of autoradiograms must be carried out in a dark room. The authors use Kodirex X-ray film and a Kodak 6B (brown) safelight. The film is obtained in 8 × 10 inch (203 × 254 mm) sheets and it can be cut to the size required in the dark room. The film must always be handled only at the edges and should not be placed on a rough surface.

A piece of X-ray film is attached to the test object using two small pieces of adhesive tape, thin layer chromatogram plates are laid carefully on to the film. The film and the object are placed between pieces of paper and then sandwiched between two glass plates to ensure good contact between them.

The sandwich is then placed in a suitable light-tight storage box; an old X-ray film or chromatography paper box will be found useful and the box is wrapped in aluminium foil to ensure it is light-tight. A dark cupboard forms a suitable place to leave the box for exposure and during this time it should not be moved.

In the dark room the X-ray film is removed from the box and separated from the object after marks have been made to facilitate relocation. The film is then developed in Kodak D19 developer for 5 min at room temperature, with constant agitation of the developing dish. The film is removed from the developer with forceps, rinsed for 10 seconds in clean water and then immersed in acid fixative and agitated frequently until it is clear (3-10 min). After washing in running water for 15-30 min the film is hung up to dry.

USING NUCLEAR STRIPPING FILM

These procedures must be carried out in a dark room under a Kodak 1 (red) safelight keeping the film away from direct light.

Nuclear stripping film (Kodak AR.10) is supplied coated on to glass plates of 120×162 mm and, since it is only 15 μm thick, it must be treated with some care. The preparation and use of this film is described below and illustrated in Figure 18.

In the dark room a scalpel is used to cut the film 1cm inside of and parallel to each of its four sides.

The central portion is then divided into eight or twelve squares of approximately equal dimensions.

After a few minutes the pieces of film will begin to separate from the glass and by sliding a scalpel between the film and the glass a piece of film may be removed. The film is lifted off the plate, turned over so that the emulsion side faces downwards and then placed on the surface of clean water at room temperature.

After leaving the film on the water surface for 3 min, during which time it swells and expands, the slide bearing the section is immersed below the water and then withdrawn so that the film drapes itself over the section and overlaps the slide along the edges. The overlapping ends are then folded to the underside of the slide and the film is straightened out with a finger and a camel hair brush. All the wrinkles must be removed. The slides are then placed in a microscope slide carrier and put into a light-proof box to dry at room temperature. When dry they are placed in a second light-a

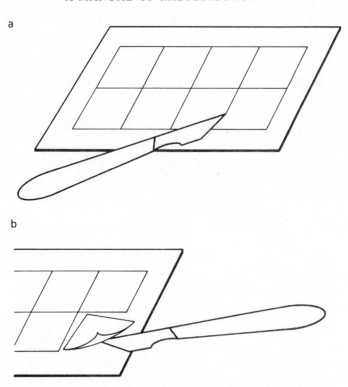

b

Turn film over before entering water bath. The emulsion side will now face down.

c

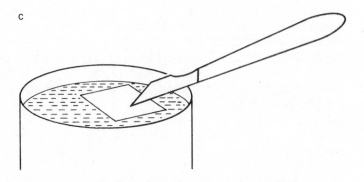

Figure 18. Method of use of autoradiographic stripping film

d

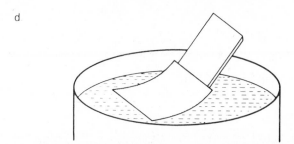

The film is straightened out with the finger and a camel hair brush.
All wrinkles must be removed.

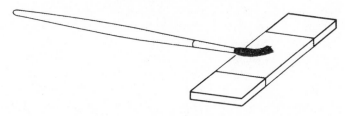

Method of use of autoradiographic
stripping film

proof box containing self-indicating silica gel and left for a suitable time to allow the specimen to irradiate the film.

When exposure is complete the rack is removed from the box and immersed in Kodak D19 developer at 18°C for 4 min with constant agitation. Development may be prolonged for up to 10 min if the image is faint.

The rack is removed and the slides rinsed twice by immersion in water and then fixed in an acid fixer for 2 min. After drying thoroughly at room temperature in a drying box the emulsion on the back of the slide is scraped off using a scalpel and the slides examined under a microscope. Specimen areas which contain isotope will appear black because of the effect of the irradiation on the overlying emulsion.

The list given below details the items necessary for preparing macroscopic autoradiograms, it is included here to avoid repetition of it in the experimental section.

Dark room, Kodak 6B safelight, Kodirex X-ray film (no screen, ester based) of size 200 × 250 mm.

Developing dishes, Kodak D19 developer and a proprietary acid fixative (or 170 g sodium thiosulphate + 25 g sodium metabisulphite made to 1 litre with water).

Forceps, 'Sellotape', scissors, light-proof box, clock, hanging rack and glass plates to act as weights.

Evaporating Solvents

The only other technique which appears in a number of experiments in this book is the evaporation of solvents in order to produce a sample suitable for counting.

Ideally solvents should be evaporated under vacuum at low temperature and a rotary evaporator is best for this. However, because such equipment is expensive it is unlikely that many establishments will be able to afford the number necessary for use with large classes.

In general, the experiments in this book require the evaporation of small volumes of solvent and the authors have found the equipment depicted in Figure 19 very suitable.

By use of a suitable aquarium pump, 'T' pieces and screw clips it is possible to have about four beakers supplied by one pump.

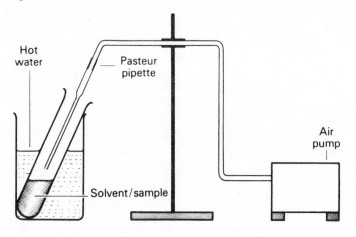

Figure 19. Simple apparatus for evaporating solvents

When evaporating ether, water is unnecessary but for higher boiling point solvents (ethyl acetate, acetone) warm water must be used. The evaporation must be carried out with great care preferably in a fume cupboard designed for inflammable solvents or in such a way that explosions initiated by naked flames or electrical contacts are avoided.

Section B

EXPERIMENTS

ENZYMIC DECARBOXYLATION OF ^{14}C-GLUTAMATE

The enzyme glutamic decarboxylase catalyses the removal of one carboxyl group from glutamate yielding γ-aminobutyrate and CO_2 as products.

$$
\begin{array}{ccc}
\text{COOH} & & \text{COOH} \\
\mid & & \mid \\
(\text{CH}_2)_2 & \longrightarrow & (\text{CH}_2)_2 \quad [+\ CO_2] \\
\mid & & \mid \\
\text{CH-NH}_2 & & \text{CH}_2 \\
\mid & & \mid \\
\text{COOH} & & \text{NH}_2 \\
\text{Glutamic acid} & & \gamma\text{-aminobutyric acid}
\end{array}
$$

The enzyme requires pyridoxal phosphate as co-factor and while it is present in many bacteria the significance of γ-aminobutyrate in bacterial metabolism is obscure.

The reaction could be followed manometrically by measuring the volume of CO_2 evolved. However, it can also be followed by measuring the amount of $^{14}CO_2$ released from universally ^{14}C-labelled glutamate over a given time. The other product of the reaction can be identified by paper chromatography and autoradiography.

Two separate experiments are described here: (1) determination of the activity of the enzyme; (2) identification of the reaction product.

The enzyme will be extracted from an acetone powder of *E. coli*. This may be obtained commercially in both crude and purified forms, either of these may be used for the first experiment but only the latter should be used in the second experiment.

EXPERIMENT 1. ACTIVITY OF GLUTAMIC DECARBOXYLASE

In this experiment the evolution of $^{14}CO_2$ from [^{14}C]-glutamate will be followed over a period of time. Four Warburg flasks will be incubated for 0, 5, 15, 30 min when the reaction will be terminated with trichloroacetic acid (TCA). The latter also helps to displace the CO_2 dissolved in the reaction fluid. The CO_2 evolved will be captured in KOH and, after dilution, can be easily counted.

Procedure

1. *Preparation of the enzyme*

Weigh out 50 mg of the *E. coli* acetone powder into a small sample bottle. Add 5 ml 0·1 M acetate buffer pH 5, mix well and stand the enzyme in ice for 5 min.

2. *Estimation*

Four Warburg flasks with double side arms will be used, label these 0, 5, 15 and 30 and their side arms 1 and 2. Lightly grease the neck joints, the rim of the centre well and the entry into the side arms of each flask by using a cotton swab. Pipette the following reagents into the chambers indicated.

Main chamber

Acetate buffer	1·0 ml
Pyridoxal phosphate	0·3 ml
[U-^{14}C]-glutamate(1μCi)	2·0 ml

Side arm 1

20% TCA	1·0 ml

Side arm 2

Enzyme extract	1·0 ml

Using a 1 ml syringe with a long needle carefully add 0·2 ml 40% KOH into the centre well of each flask. Equilibrate the flasks either in an incubator or in a water bath at 30°C for 5 min (the water in the bath only being deep enough to cover the base of the flask to a depth of 0·5-1 cm). Stopper the flasks and tip the enzyme into the main chamber with gentle swirling.

At 0, 5, 15 and 30 min tip the TCA from side arm 1 into the main chambers of the respective flasks, mix by gentle swirling and allow them to stand for at least another 30 min.

Carefully pipette 0·1 ml of the KOH solution from each centre well into correspondingly labelled tubes and dilute them to 1 ml with water. Add one drop diluted detergent to each tube and mix.

Using duplicate planchets for each tube (i.e. eight in all) add 0·1 ml diluted sample to respective planchets, add water dropwise until the base of each planchet is covered and then evaporate each carefully to dryness.

If a scintillation counter is being used matters can be simplified if only 0·1 ml of KOH is used in the centre well and a rolled 2 cm diameter glass fibre disc is inserted into the wells. This can be removed, placed in the bottom of the scintillation vial and counted after covering with 3 or 4 ml of non-aqueous scintillant.

Counting and expression of results

Count each planchet or scintillation bottle at least three times for a long time period on each count. Average the six counts for each sample and subtract the average background count.

Plot corrected count vs. incubation time and determine the rate of reaction in counts per mg powder per hour from the graph.

Notes on the method

The acetone powder used here contains enzymes other than glutamic decarboxylase. Also terminating the reaction with TCA destroys any remaining glutamate and any γ-aminobutyrate formed. For these reasons, it is not possible to determine the amount of γ-aminobutyrate formed in these reactions.

EXPERIMENT 2.   IDENTIFICATION OF THE DECARBOXYLATION PRODUCT

In this experiment a purified sample of glutamic decarboxylase must be used. The experimental procedure is modified to preserve any γ-aminobutyrate formed in the reaction, and this may then be separated quite easily from glutamate by paper chromatography. Autoradiography or elution and counting can be used to determine if the product is radioactive.

In this case, a single Warburg flask is used, the CO_2 evolved is again captured in KOH, but the reaction is terminated by heating to 60°C. This also helps to expel the CO_2 dissolved in the reaction mixture.

Procedure

1. *Extraction of the enzyme*
Weigh the amount of acetone powder corresponding to 2 enzyme units and dissolve this in water. Stand the solution in ice until required.

2. *Incubation*
Prepare a single side arm Warburg flask as in the previous experiment. Into the main chamber, pipette the following:

Acetate buffer	1 ml
Pyridoxal phosphate	2 ml
[^{14}C]-glutamate	$1\mu\,C_i$

Into the side arm, pipette 1 ml of enzyme extract and into the centre well, 0·2 ml 40% KOH. Equilibrate the flask in an incubator or water bath at 30°C for 5 min. Stopper the flask and side arm vent, tip the enzyme into the main chamber, mix by swirling and incubate for 30 min. To stop the reaction place the flask in a water bath at 60°C for 15 min.

Carefully remove 0·1 ml of the contents from the centre well into a sample tube and retain it for counting. Remove the remaining KOH with a Pasteur pipette, wash the centre well twice with a few drops of water and discard the KOH and washings to waste. Carefully transfer the contents of the main chamber to a small flask using a Pasteur pipette and evaporate under reduced pressure at 50-60°C.

Dissolve the contents of the flask in 0·2 ml water and spot 10 μl of this onto a 30 × 10 cm chromatography paper. Spot also 10 μl each of glutamate, γ-aminobutyrate and pyridoxal phosphate as control materials. Develop the paper by overnight ascending chromatography in butanol/acetic acid/water, dry it and spray with ninhydrin. Visualise the colours by heating in an oven at 80-100°C until the spots appear; mark around them with a pencil.

Prepare an autoradiogram of this paper as described earlier (Chapter 7) allowing seven days exposure. Discuss the results obtained.

Counting and expression of results

The KOH sample is counted as detailed in the previous experiment. Average the sample counts and subtract the average background.

A standard planchet or vial containing a known activity of [U-^{14}C] glutamate will be supplied and this should be counted in the same way as the samples. Use the result to calculate the absolute activity of the $^{14}CO_2$ released and hence the amount of glutamate decarboxylated.

Notes and suggestions for further work

The amount of γ-aminobutyrate formed in this experiment is fairly small but can be detected with the very sensitive ninhydrin reagent. It is also readily detectable on the autoradiogram. Pyridoxal phosphate shows up on the chromatogram as a large yellow spot between glutamate and γ-aminobutyrate.

The experiment may be extended by running a second spot from the reaction and extracting the glutamate and γ-aminobutyrate from the paper and counting them.

If a paper is developed in ethanol/ammonia/water (90:10:10) γ-aminobutyrate may be separated from α-aminobutyrate which is another possible product.

Student level
HNC students and above.

Time required

Experiment 1	3-4 h
Experiment 2	basic experiment 3-4 h
	chromatogram development 12-18 h
	autoradiogram preparation 30 min
	exposure 1-2 wk
	development 30 min

Materials required

Experiment 1

Per group

Acetone powder *E. coli* Type 1 Sigma G2002	50 mg	Specimen tubes	4
0·1 M Acetate buffer pH 5: 0·6 g glacial acetic acid + 80 ml water adjusted to pH 5 and diluted to 100 ml	10 ml	Syringe with long needle, 1 ml	1
Pyridoxal phosphate 100 μg ml^{-1} in acetate buffer	2 ml	*For general use* Ice and ice baths Pipettes—various capacities, graduated	
[U-^{14}C]-glutamate (1 mg ml^{-1}; 1 μCi 2 ml^{-1}	8 ml	Grease	
TCA, 20%	10 ml	Water baths at 30°C	
KOH, 40%	2 ml	Dilute detergent	
Warburg flasks (double side arm) and stoppers	4	Planchets (and hot plates or heat lamps) or scintillation vials (and scintillant)	
Clock	1	Pasteur pipettes and teats	
		Cotton swabs	

Experiment 2

Per group

Acetone powder *E. coli* (purified) Type 2 Sigma G2126	2 units	Pasteur pipettes and teats
0·1 M acetate buffer	2 ml	Planchets or scintillation vials
^{14}C-glutamate (1 mg ml^{-1}; 1 μCi)	1 ml	Hot plates and heat lamps Vacuum pump and pressure tubing Chromatography paper 10 × 30 cm
Pyridoxal phosphate 100 μg ml^{-1} in acetate buffer	2 ml	Whatman No. 1
KOH, 40%	0·5 ml	Chromatography tanks for ascending chromatography with butanol/
Sample tube	1	acetic acid/water (4:1:5) top
Warburg flask (single side arm) and stoppers	1	phase Chromatography controls—gluta-
Clock	1	mate and γ-aminobutyrate (8 mg
1 ml syringe with long needle	1	ml^{-1}) Sprays of ninhydrin in acetone
50 ml round bottom flask	1	Microlitre pipettes
For general use		Oven at 80°-100°C
Hair dryer		Materials and facilities for auto-
Grease		radiography
Pipettes—graduated		Standard planchet or vial contain-
Water baths at 30° and 60°C		ing 0·01 μCi[^{14}C]-glutamate
1% detergent		Ice Waste bottle

BIOSYNTHESIS OF GLUTAMINE FROM [U-¹⁴C]-GLUTAMATE

$$
\begin{array}{ccccc}
& & & & \begin{array}{l} COOH \\ | \\ CH.NH_2 \\ | \quad (4) \\ (CH_2)_2 \\ | \\ CONH.OH \end{array} \\
\begin{array}{l} COOH \\ | \\ CH.NH_2 \\ | \\ (CH_2)_2 \\ | \\ COOH \\ (1) \end{array} &
\begin{array}{c} ATP \\ \diagdown \; ADP \\ \longrightarrow \end{array} &
\begin{array}{l} COOH \\ | \\ CH.NH_2 \\ | \\ (CH_2)_2 \\ | \\ COO\;\textcircled{P} \\ (2) \end{array} &
\begin{array}{c} NH_2OH \\ \nearrow \\ \searrow \\ NH_3 \end{array} &
\begin{array}{l} COOH \\ | \\ CH.NH_2 \\ | \quad (3) \\ (CH_2)_2 \\ | \\ CONH_2 \end{array}
\end{array}
$$

The basic amino acid glutamine (3) is an extremely important source of amine groups in transamination reactions and hence its biosynthesis is of interest.

In the cell it can be formed by the amination of glutamate (1), the reaction involving ATP and the enzyme glutamine synthetase. The latter catalyses a two-step reaction
(a) the activation of glutamate with ATP giving the phosphate (2)
(b) amination of the latter to give glutamine (3).

The amine group is normally supplied by ammonia. However, it is known that hydroxylamine (NH_2OH) may also be utilised in which case the product is glutamyl hydroxamate (4). The latter can easily be separated from glutamate by chromatography.

In this experiment, the biosynthesis of glutamyl hydroxamate by an acetone powder of peas will be investigated. The starting material will be [U-¹⁴C]-glutamate and the glutamyl hydroxamate formed will be estimated by autoradiography.

Procedure

1. *Preparation of the acetone powder*

Remove the testas from about 15–20 g germinated peas and chill to 0°C. Using an ice-cold mortar and pestle, grind the peas to a paste with the *minimum* of lumps. Add 100 ml acetone at −20°C to the paste and mix gently, then add a further 100 ml cold acetone and mix again. Filter through a sintered glass funnel under vacuum leaving as many lumps as possible in the mortar. Regrind these lumps, then add 50 ml cold acetone, mix and filter. Avoid drawing air through the material on the filter as this tends to denature the enzyme.

Wash the finely ground material on the filter with 50 ml cold acetone followed by 50 ml ether at −20°C. Turn the material out on to a paper towel or filter paper and allow it to dry. To assist in drying move the material about with a spatula from time to time.

When dry, finely sieve the powder and discard any lumps.

2. *Extraction of the enzymes*

Weigh out 2–3 g of the fine powder into a boiling tube. Cool to 0°C in ice/water. Add 20 ml ice-cold 0·1 M NaHCO₃ and mix with a glass rod to give an even suspension. Leave in the ice bath for about 10 min, shaking occasionally. Centrifuge the suspension in ice-cold polythene tubes at 2,000–3,000 g for 10 min. Remove the supernatant with a Pasteur pipette, recentrifuge if necessary and keep cold in ice.

3. *Assay procedure*

Set up the following solutions (ml) in each of two test tubes. A tube containing 2 μCi of [U-^{14}C]-glutamic acid in 0·5 ml water may be supplied in which case this should be used as tube 1.

Solutions	Tubes	
	1	*2*
Tris-HCl pH 7·5	0·5	0·5
ATP	1·0	1·0
[U-^{14}C]-glutamate	0·5	0

Glutamate	0·5	0·5
MgSO$_4$	0·1	0·1
NH$_2$OH.HCl	0·1	0·1
Water	0·8	2·3

At zero time add 1 ml enzyme extract to tube 1 and incubate this and the control tube (2) at 30°C.

During this time, prepare a Whatman number 1 chromatography paper (about 10 × 40 cm) for descending chromatography, folding it to fit the equipment available. The solvent must be able to run for at least 35 cm. After 15, 30 and 45 min incubation remove 10 μl from tube 1 and spot this on to the paper. Spot also 10 μl of the glutamate and glutamyl hydroxamate standard solutions provided and 10 μl of the contents of tube 2 after 45 min incubation as a non-enzymic control.

Develop the paper for 18-24 hours in the butanol/acetic acid/water solvent and during this time the solvent should be allowed to drip off the bottom of the paper in order that the spots will move the maximum distance and give the best separation possible. Remove the paper from the tank and dry it in a fume cupboard. Spray with ninhydrin and heat in an oven at 120°C until the spots appear. Draw a pencil ring around the spots and make a drawing of the paper for inclusion in your report.

Prepare an autoradiogram of the paper ensuring that you are able to relocate the chromatogram on the film and leave it to expose for 14 days before developing the film as indicated in Chapter 7. Compare the paper .and the autoradiogram.

Student level
Sixth formers to degree students.

Time required
Experiment 3-4 h
Chromatogram development overnight
Chromatogram assessment and autoradiogram preparation 1 h
Autoradiogram exposure 14 d
Autoradiogram development 1 h

Materials required

	Per group
Germinating peas—soaked in water overnight and kept at 0°C for one hour before the experiment	15-20 g
Acetone @ −20°C	300 ml
Ether @ −20°C	50 ml
$NaHCO_3$, 0·1 M	25 ml
Tris-HCl, 0·4 M pH 7·5	2 ml
ATP, 0·5 M	2 ml
Glutamate, 0·5 M	1 ml
[U-^{14}C]-glutamate (2 μCi) possibly dispensed into a test tube	0·5 ml
$MgSO_4$, 1M	1 ml
NH_2OH. HCl, 1 M	1 ml
Small beaker	1
Large mortar and pestle (ice-cold)	1
Sintered glass funnel, Buchner flask and pump	1
Boiling tube	1
Glass rod	1
Clock	1
Microlitre pipettes	5
Test tubes and rack	6
Whatman Number 1 chromatography paper 45 × 10 cm or longer	1
Spectrophotometer and cuvettes	1
Measuring cylinder (100 ml)	1

For general use

Ice and ice baths
Filter papers or paper towels
Fine wire or plastic sieves
Balances (not necessarily analytical)
Bench centrifuges and tubes
Pasteur pipettes and teats
Water baths at 30°
Glutamate and glutamyl hydroxamate chromatography standards 2 mg ml—^{11} } 0·5 ml
Butanol/acetic acid/water (4:1:5)—top phase
Chromatography tanks and accessories for descending chromatography
Hair dryers
Materials for autoradiography (see Chapter 7)
Oven at 100°C
Sprays containing 0·5% ninhydrin in acetone
Pipettes—various capacities, graduated

BIOSYNTHESIS OF PLANT LIPIDS FROM [U-^{14}C]-ACETATE

Two of the important groups of lipids found in plants are:
(a) triglycerides; non-polar lipids found in globules in the cytoplasm and acting as storage products
(b) phospholipids; polar lipids found primarily in membranes and acting mainly as structural components.

All lipids in these groups are syntheised from acetate. Germinating peas carry out a number of intense metabolic activities including lipid synthesis, and slices of such tissue will efficiently incorporate radioactive carbon from acetate. Normally, germinating peas synthesise lipid from starch storage reserves and the amount synthesised is small but can, however, be easily demonstrated using radioactive tracers.

Procedure

1. *Preparation of the pea slices*

For each student group plant 10–12 peas about 25 mm deep in vermiculite, water them but do not let them become saturated. After 3–4 days the peas will have germinated sufficiently to use.

Break off and discard the radicals, carefully remove the testas and after washing the cotyledons in phosphate buffer, cut them into thin slices (about 0·5 mm) using a razor blade. Collect the slices in phosphate buffer, strain them through muslin in a filter funnel and wash the slices on the muslin with a little more buffer. Do not let the slices dry out.

2. *Incubation*

You will be provided with a 25 ml conical flask containing 5 ml phosphate buffer and 5 μC_i[U-^{14}C]-sodium acetate. Into this flask weigh about 1 g of the thinnest pea slices, stopper the flask and incubate it at room temperature for as long as possible in a *fume cupboard* with occasional shaking.

Stop the reaction by adding 0·1 ml 18N H_2SO_4. After a few minutes decant the supernatant and wash the slices with 2 ml portions of buffer to remove excess activity. Combine the supernatant and washings and discard as radioactive liquid waste. To the tissue slices add 10 ml 2:1 chloroform/methanol, stopper and leave for as long as possible preferably with occasional shaking.

The experiment may be left until the next laboratory period at this stage, but if this is intended it is advisable to flush out the flask with nitrogen, stopper tightly and store at 0–4°C. This will prevent lipid oxidation and subsequent problems with the chromatographic separation.

Decant the solvent from the slices into a boiling tube.

Wash the slices with 5 ml solvent in two portions and decant this into the boiling tube. Discard the cotyledons as solid radioactive waste.

Add a small amount of anhydrous Na_2SO_4 to the tube and rotate gently to remove the water. If necessary add further small portions of Na_2SO_4 until the solvent is clear. Filter through a paper into a clean boiling tube and wash the paper with a *small amount* of chloroform/methanol. Alternatively, filter the extract without preliminary drying through a phase separating paper e.g. Whatman 1 PS. Evaporate the filtrate to dryness by removing the solvent in a stream of nitrogen or air.

3. *Thin layer chromatography of the lipid extract*

You will be provided with a 20 × 20 cm plate spread 0·25 mm thick with kieselgel G and also with samples of known lipids, possibly a triglyceride, cholesterol, cholesterol oleate, a phospholipid and a fatty acid.

Dissolve the lipid extract in a few drops of chloroform and apply the solution to the plate as indicated in Figure 20. The sample spot should be about 5-10 µl and the remainder of the extract is applied as a streak. Apply 10 µl of the controls as

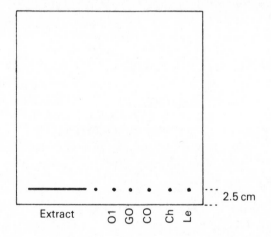

Figure 20. Thin layer plate for lipid extract (O1-oleate, GO-glyceryl trioleate, CO-cholesterol oleate, Ch-cholesterol, Le-lecithin)

indicated in the diagram. Develop the plate in the petroleum ether/ether/formic acid solvent and dry it at room temperature before exposing it to iodine vapour in a chromatography tank. Do not expose for longer than is necessary to visualise the spots. Make a drawing of the plate, indicating the relative intensities of the spots and with a sharp pencil gently ring round the spots.

Prepare an autoradiogram of the sample spot and streak, and leave it to expose for 2 weeks before developing the film. When the film is dry locate it on the tlc plate and determine which bands or spots are radioactive. Compare the relative activities and relative colour intensities of the lipids present in the extract; particularly the phospholipids and triglycerides.

Scrape the kieselgel containing the major radioactive bands from the plate into separate clean tubes. Add 1-2 ml chloroform to each and leave for 10-15 min before filtering through a Pasteur pipette plugged with cotton wool as indicated in Figure 21. Use a fresh Pasteur pipette for each band. Evaporate the solvent in a stream of air or nitrogen (see Chapter 7) and when dry either add scintillation fluid and count by scintillation counting or plate the extracts on to planchets and count on an end window counter.

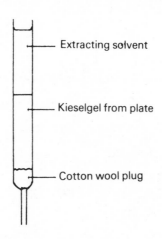

Extracting solvent

Kieselgel from plate

Cotton wool plug

Figure 21. Kieselgel extraction

The autoradiographic step could be by-passed if desired and the lipid samples which are visualised by iodine vapour scraped off and eluted as described above.

4. Counting the samples

If Geiger-Muller counting is being used count all the samples three times each for a long time period (say 300 sec) on each count. Determine the background similarly. Average the counts for each sample and subtract the average background count. Assume that the total activity is 100% and determine the relative activity of each band.

Suggestions for further work

The polar lipids remain at the origin in this solvent system but can be separated from each other using tlc and a chloroform/methanol/water/formic acid solvent system.

Student level

Sixth formers and above.

Time required

Experiment 2–4 h
Chromatographic preparation and development 2 h
Sample extraction and counting 1 h
Optional autoradiography 14 d

Materials required

	Per group	For general use
Phosphate buffer (0·5 M) pH 7	100 ml	Razor blades
Chloroform/methanol 2:1	25 ml	Small beakers Muslin
12 peas germinated for 72 hours in vermiculite		Filter funnels and papers Balance
25 ml conical flask + 5 ml phosphate buffer + 5 μCi [U^{14}C]-acetate and bung to fit	1	Pasteur pipettes and teats Boiling tubes Test tubes and racks Microlitre pipettes (10 μl)
20 × 20 cm tlc plate spread 0·25 mm thick with kieselgel G	1	Pipettes various capacities, graduated Chromatography tanks

Planchets or scintillation vials	solid waste
Heat lamps or scintillation fluid	18N H_2SO_4 (con. H_2SO_4:water, 1:1)
Equipment for removing solvents (see Figure 19)	Dilute detergent
	Anhydrous Na_2SO_4 or Whatman 1
Absorbent cotton wool	PS papers
Bottle for radioactive waste	Bottles of chloroform
Polythene bags or pedal bin for	Solid iodine

Glyceryl trioleate	or other
Cholesterol	standard
Cholesterol oleate	lipids 1 M
Lecithin	in $CHCl_3$/0·1%
Oleic acid	hydroquinone

ESTIMATION OF THE GLUTAMATE CONTENT OF GLUTATHIONE

Glutathione is very widely distributed throughout the plant and animal kingdom. It has been shown to be a tripeptide consisting of glutamate, cysteine and glycine and has the following structure:

$$HOOC.CH.CH_2.CO \;.NH.CH.CO. \;NH.CH_2.COOH$$

$$\underset{\text{Glutamic acid}}{NH_2} \qquad \underset{\text{Cysteine}}{CH_2.SH} \qquad \text{Glycine}$$

Glutathione is an important reducing compound being associated in particular with the maintenance of the reduced state of sulphydryl (-SH) groups.

Being a peptide, glutathione can be hydrolysed by acid but during this procedure the cysteine is partially oxidised to cysteic acid. To avoid obtaining a mixture of these two compounds the glutathione is first oxidised with performic acid to convert all the cysteine to cysteic acid. By adding [14C]-glutamate of known specific activity to the hydrolysis mixture and then re-isolating the glutamate by paper chromatography and redetermining the specific activity, it is possible to find the glutamate content of the glutathione.

Since this value is known it can be used as a check on the accuracy of the method which is capable of giving an answer within 1% of the theoretical value.

Procedure

1. *Preparation of performic acid*
To 9 ml formic acid add 1 ml 100 volume H_2O_2 and allow this to stand for 30 min at room temperature before use.

2. *Preparation of a standard paper*
Into a 50 ml round bottom flask weigh out about 10 mg cysteine, add 0·2 ml performic acid and allow to stand at room temperature for 20 min. Evaporate to dryness either on a rotary evaporator or in a desiccator under vacuum.

Add 0·5 ml water and re-evaporate, add 0·8 ml water and evaporate again; this will remove excess performic acid. Dissolve the cysteic acid so formed in 1 ml water and spot 10 μl of this on to a Whatman Number 1 chromatography paper at least 45 cm long, folded to suit the apparatus available. Spot also 10 μl of the glutamate, glycine and other authentic solutions provided. Run the paper in butanol/acetic acid/water by descending chromatography for at least 18 h during which time the solvent should be allowed to drip off the bottom of the paper. Dry the paper in a fume cupboard, spray it with ninhydrin and heat in an oven at 100°C until the spots appear. From the paper determine the order and relative positions of the compounds.

3. *Hydrolysis of glutathione*
Into a 50 ml round bottom flask weigh accurately about 10 mg glutathione and treat it with performic acid exactly as above. Dissolve the dried material in 1 ml water and using a Pasteur pipette transfer it to a Carius tube. Wash the flask with two 0·5 ml portions of concentrated HCl and add these to the tube. Seal the tube and heat at 110°C for 18 h. After cooling carefully open the tube and quantitatively transfer the contents to a 50 ml round bottom flask by washing the tube well with water. Evaporate to dryness as before.

4. *Isotope dilution*

Weigh out accurately about 10 mg glutamate and dissolve it in 100 ml water in a volumetric flask. Pipette 1 ml of this solution into a test tube and add 5 μCi of high specific activity [^{14}C]-glutamate. After mixing, use a 0·1 ml pipette to transfer 0·05 ml of this solution to each of two planchets, add 1 drop detergent to each and dry under a heat lamp. Retain the planchets for counting.

Add the remaining 0·9 ml of radioactive glutamate to the hydrolysed glutathione, washing both the tube and the pipette used for the above transfers with water and adding the washings to the hydrolysate. Evaporate to dryness as above.

Dissolve the hydrolysate in 1 ml water and spot it on to four 10×45 cm Whatman Number 1 chromatography papers putting 5 spots of 10 μl on to each paper. Develop the papers in the solvent described in section 2. After drying, cut a 1·5 cm strip off both sides of each paper, marking the strips clearly so that they may be relocated against the correct paper. Visualise the amino acids on the strips with ninhydrin, as before. Relocate the strips alongside the papers and mark the position of the glutamate band on the unsprayed portion of each paper. Cut out these portions of the papers but avoid touching them with the fingers, then cut each one into small pieces and place these in a small flask. Elute the paper pieces with two 4 ml portions of water, combine the washings and evaporate them almost to dryness. Plate the extract on to a planchet using as little heat as possible as the glutamate must be recovered from the planchet at a later stage.

After counting place the planchet into a small beaker, add 2 ml water and heat over a small bunsen flame. Using a Pasteur pipette transfer the solution to a 5 ml volumetric flask then repeat with a further 2 ml portion of water. Make up to 5 ml.

Determine the glutamate concentration of this solution by treating duplicate 1 ml samples with ninhydrin in the manner described in section 5.

5. *Calibration curve*

Using the remaining glutamate solution prepared in

section 2, make a series of dilutions ranging from 1×10^{-2} to 1×10^{-3} mg ml^{-1}. Treat duplicate 1 ml samples of these with 2 ml ninhydrin reagent and heat in a boiling water bath for 15 min. Cool and after adding 5 ml 50% alcohol to each tube determine the absorbance at 420 nm using a blank of 1 ml water which has been treated in the same way. Plot absorbance against glutamate concentration mg ml^{-1}.

Counting and calculation

Determine the activity of each planchet by counting it three times for 100 seconds each count, average the counts for each and subtract the average background.

From the calibration curve determine the amount of glutamate present in the 5 ml of extract from the papers. If high specific activity [14C]-glutamate has been used the weight of radioactive glutamate present can be assumed to be negligible and the amount added to the hydrolysate will be that present in the solution originally prepared i.e. 0·09 mg.

From the above data determine the specific activity of both samples of glutamate, in this case as counts mg^{-1}. Then

$$\frac{\text{Weight unlabelled material}}{\text{Weight labelled}} = \left(\frac{\text{Specific activity additive}}{\text{Specific activity isolate}} - 1 \right)$$

Note

If scintillation counting is to be used in this experiment the only change required is in section 4 where it is possible to remove half of the elution volume for counting in a suitable scintillation mixture and assaying the remainder for glutamate concentration. The appropriate volume corrections are easily made.

Student level
HNC and degree students.

Time required
Preparation of hydrolysis mixture 2 h
Preparation of standard chromatogram and calibration curve 2 h

Hydrolysis overnight
Preparation of hydrolysate chromatogram 2 h
Development of chromatogram 18–24 h
Counting of radioactive samples 2 h

Materials required

	Per group	
High specific activity [U-^{14}C]-glutamate	5 μCi	Spectrophotometer and cuvettes 1
Butanol/acetic acid/water (100:22:50)	100 ml	Water bath at 100°C 1
Ninhydrin reagent (prepared fresh before use)		*For general use*
Dissolve 1·5 g ninhydrin and 0·05 g stannous chloride in 50 ml methyl cellosolve (ethylene glycol monomethylether). Filter. Add to 50 ml acetate buffer pH 5	20 ml	Formic acid
		Hydrogen peroxide, 100 volume
		Cysteine
		Glutamate, glycine and other authentic compounds, 0·5 M in water
		Glutathione AR
		Glutamate AR
Alcohol 50%	50 ml	Clocks
Small Carius tube about 10 ml capacity	1	Balances
25 ml conical flask	1	Pipettes—various capacities
10 ml measuring cylinders	2	Rotary evaporator or vacuum desiccator with P$_4$O$_{10}$
100 ml volumetric flask	1	Vacuum pump
5 ml volumetric flask	1	Concentrated HCl
50 ml round bottom flasks	5	Ninhyrin in acetone, 0·5%
10 ml beaker	1	Chromatography sprayers
0·1 ml pipette	1	Microlitre pipettes
Test tubes in rack	1	Ovens at 100° and 110°C
Whatman Number 1 chromatography papers 55 × 10 cm	5	Pasteur pipettes and teats
		Bunsen, tripod and gauze
		Planchets, (and heat lamps or hot plate) or scintillation vials (and scintillant)
Tank for running the above by descending method	1	Dilute detergent
		Liquid waste bottle

THE PHOSPHATE CONTENT OF BONE

Although bone is a living material the greater part of it consists of the mineral calcium phosphate and it is possible to estimate the phosphate content using the isotope dilution method.

If bone is digested in concentrated nitric acid, the phosphate is solubilised as orthophosphate. If a known quantity of $^{32}PO_4^{3-}$ of known specific activity is then added to the digestion mixture it will become intimately mixed with the unlabelled phosphate from the bone. Utilising the standard techniques of gravimetric analysis the phosphate may then be precipitated as $MgNH_4PO_4$ which, after drying under specified conditions, has the composition $MgNH_4PO_4.6H_2O$. Thus the phosphate content of the precipitate is known and from the specific activity of this the dilution of the radioactivity can be determined.

This method is useful since all the phosphate in the mixture need not be precipitated, although the final precipitate must be pure. Since the phosphate content of a bone is likely to vary in different regions *either* a number of determinations on different samples should be done and the mean obtained *or* a homogeneous powder should be produced.

Procedure

Carry out the following procedure in duplicate.

Into a labelled 10 ml conical flask weigh out accurately about 0·1 g of the bone fragments provided. Note the weight taken.

Add 0·5 ml conc. HNO_3 and heat *gently* on a hot plate in a fume cupboard until the bone dissolves, then evaporate *almost* to dryness.

To the flask add 2·5 ml water and transfer the contents to a test tube rinsing the flask with a further 2·5 ml water. Add to the tube, as accurately as you can, 1 ml of the $^{32}PO_4^{3-}$ solution provided (about 1 μCi ml^{-1}). Add 3-4 drops concentrated HCl, 3-4 drops methyl red and 4 ml magnesia reagent. Add concentrated NH_4OH dropwise and with stirring until the solution goes yellow, then add 2 drops more. Set the tube aside for 10 min.

Pour the contents of the tube into a labelled centrifuge tube and centrifuge at low speed for a few minutes to sediment the precipitate. Decant off the supernatant to the radioactive waste bottle. Resuspend the precipitate in 4-5 ml of $1 \cdot 5$ M NH_4OH and re-centrifuge. Repeat this washing a further three times decanting the supernatant into the waste bottle on each occasion.

Finally, resuspend the precipitate in a few drops of $1 \cdot 5$ M NH_4OH and add a few drops of this to a *weighed* planchet. Add 1-2 drops of dilute detergent and gently swirl the planchet until the material covers the bottom. Evaporate *gently* to dryness, cool and re-weigh.

Prepare two standard planchets by evaporating $0 \cdot 3$ ml of the original $^{32}PO_4^{3-}$ solution and 1 drop detergent on to planchets. Count all the samples twice for 100 sec on each occasion.

Calculation

The precipitate obtained is $MgNH_4PO_4.6H_2O$ and this has 39% PO_4^{3-} by weight. Determine the specific activity (Z) of the experimental samples by taking the count rate obtained and its known weight and calculating counts mg^{-1} PO_4^{3-} in the precipitate.

Determine also the specific activity of the standards (Y) from the count rates and the known phosphate concentration of the material.

Weight unlabelled material (X) =

$$\text{weight labelled material}\left(\frac{\text{specific activity additive (Y)}}{\text{specific activity isolate (Z)}} - 1\right)$$

Determine X and from it the % PO_4^{3-} in your sample.

Student level
HNC to degree level.

Time required
2–3 h

Materials required

	Per group		
10 ml conical flasks	2	$^{32}PO_4^{3-}$ 1 μCi ml^{-1} and of	
Test tubes in rack	6	known weight	3 ml
Polythene centrifuge tubes	2	*For general use*	
Stirring rods	1	Concentrated HCl	
Heat lamp or hot plate	1	Methyl red	
Measuring cylinders 5 ml	1	Planchets	
Clock	1	Balances	
NH$_4$OH (0·880 s.g.)	3 ml	Bench centrifuges	
NH$_4$OH, 1·5 M	20 ml	Syringes, 1 ml	
Magnesia reagent—dissolve 25 g MgCl$_2$.6H$_2$O and 50 g NH$_4$Cl in 250 ml water. Add excess NH$_4$OH, filter. Make acid with dilute HCl. Dilute to 500 ml		Test tube holders	
		Pipettes, 5 ml graduated	
		Pasteur pipettes and teats	
		Waste bottles	
		1% detergent	
		Bone fragments	
	8 ml	Concentrated HNO$_3$	

THE SYNTHESIS OF PROTEIN FROM [^{3}H]-L-ALANINE IN BARLEY SEEDLINGS

The synthesis of protein is one of the most fundamental and interesting of biological processes and obviously it is desirable to be able to demonstrate it, and if possible to examine some of the factors effecting it.

In this experiment, young barley seedlings are incubated with a radioactive amino acid and protein is isolated from a seedling homogenate by acetone precipitation. It is probable that most amino acids could be used in the experiment; the authors use L-alanine because it is available within the department. Tritiated amino acids have the advantage of cheapness and not producing a radioactive gas by-product. They require however a scintillation counter for their estimation.

Procedure

Weigh out approximately 20 g of barley seedlings into a 250 ml conical flask and add 50-100 ml of a suitable nutrient solution containing 10 mg of casein hydrolysate and 3-5 μCi of a radioactive amino acid. Illuminate overnight at room temperature either in a shaking water bath or with aeration. A fume cupboard or CO_2 trap will be necessary if a ^{14}C labelled amino acid is used.

Collect the seedlings by filtration through glass wool and wash them with 100 ml of water, followed by 50 ml of 1% amino acid solution. Use forceps to transfer the seedlings to a small beaker and use sharp pointed scissors to cut them into small pieces.

Transfer the pieces to the cup of a homogeniser (the MSE top-drive homogeniser with a 100 ml cup has proved satisfactory) and add 20-30 ml of water. Seal any opening at the top of the homogeniser cup with Parafilm and homogenise the seedling pieces.

Filter the homogenate through glass wool into a small beaker and centrifuge at full speed on a bench centrifuge for five minutes to remove small tissue and cell fragments.

Measure the volume of the supernatant and remove 1 ml to a scintillation vial or aluminium planchet. Use a measuring cylinder to add two volumes of acetone to the remaining supernatant and then allow the solution to stand for 15 minutes. Recentrifuge, remove the supernatant and transfer 1 ml to a scintillation vial or aluminium planchet.

Resuspend the precipitate *very thoroughly* in 20 ml ammonium sulphate using a glass rod and vortex mixer or if possible a Potter-Elvejhem homogeniser. Centrifuge the suspension at full speed for 3 min to remove particulate material and again measure the volume of the supernatant.

Add two volumes of acetone and allow to stand for 15 min. Centrifuge for 5 min, remove the supernatant and transfer 1 ml to a scintillation vial or aluminium planchet. Resuspend the precipitate in 2 ml of ammonium sulphate and transfer 1 ml to a scintillation vial or planchet.

Add 9 ml of a dioxan or triton based scintillant or dry the samples onto their planchets with a heat lamp.

Counting and expression of results

Count each sample twice and determine the mean count rates. Obtain a background count rate using a vial of scintillant or an empty planchet and subtract this value from that for each experimental sample.

The centrifuged seedling homogenate should produce a high count rate and the two supernatants from the acetone precipitations substantially lower ones. The resuspended final precipitate should produce a count markedly higher than its supernatant. The fact that this precipitate contains a significant quantity of protein can easily be demonstrated with standard protein colour reactions.

Student level
Sixth formers to degree students.

Time required
Seedling incubation overnight
Preparation of extract 2 h

Materials required

	Per group		
Barley seeds germinated in damp vermiculite (preferably at 30°C) for about 4-5 days	25 seeds	L-amino acid solution 1%	50 ml
		Acetone (preferably at 0°C)	100 ml
A simple nutrient solution such as:	100 ml	0·1 M Ammonium sulphate	25 ml
KH$_2$PO$_4$ 10 mg		Glass rod or Potter-Elvejhem homogeniser	1
KNO$_3$ 18 mg	added to 100 ml tap water	Vortex mixer	1
MgSO$_4$,7H$_2$O 10 mg		Pipette, 1 ml	1
Ca(NO$_3$)$_2$.4H$_2$O 15 mg		Scintillation vials (and water absorbing scintillant) or planchets (and a heat lamp)	5
casein hydrolysate 10 mg			
radioactive amino acid 3-5 μCi		Glass wool	
500 ml conical flask	2	Large filter funnel	1
100 ml and 25 ml measuring cylinder	1 each	Forceps and pointed scissors	1
100 ml beaker	1		

For general use

Balances

Shaking water baths or air pumps and spargers

Mechanical homogenisers

Bench centrifuges and 50 ml tubes

Parafilm

THE PHOTOSYNTHETIC SYNTHESIS OF AT^{32}P

It has been known for some time that photosynthesis consists of two processes:

(*a*) a light-dependent one in which light is used to generate chemical energy in the form of ATP and reducing power in the form of NADPH

(*b*) a light-independent one in which these components are used to convert CO_2 into a number of useful organic compounds.

This experiment is connected with the first of these two processes. The light energy is trapped by chlorophyll and then used in the phosphorylation of ADP to ATP. Thus if $^{32}PO_4^{3-}$ is supplied some of this will be incorporated into the ATP produced and this can be estimated by determining the level of radioactivity. The AT^{32}P may be separated from the unused $^{32}PO_4^{3-}$ by solvent extraction, the latter being made soluble in an organic solvent.

Procedure

The experiment can be divided into six stages; isolation of the chloroplasts, determination of chlorophyll concentration, demonstration of the ability to phosphorylate, the experimental synthesis of AT^{32}P, isolation of AT^{32}P and finally the counting procedure.

Chloroplast isolation

Homogenise 30 g of leaf material in 100 ml of cold homogenising medium by macerating for 5 sec at full speed in a cold bottom drive homogeniser. Filter the product through multiple layers of cotton organdie and nylon bolting cloth (25 μm pore). Centrifuge the filtrate at 3,000 g for 60 sec at 0-4°C and resuspend the pellet in 2 ml resuspension medium. Use cold solutions and apparatus throughout and keep the isolated plastids cold by immersion in ice. It may be best if a single large preparation for the whole class is made.

Determination of chlorophyll concentration

It is desirable to know the concentration of chlorophyll in the suspension so that a suitable aliquot can be used in the experiment. Chlorophyll concentration can be determined from its light absorption.

Prepare a solution of 0·2 ml chloroplast suspension in 10 ml of 80% acetone. After filtration measure the absorption at 645 nm and 663 nm against a blank of 80% acetone. The chlorophyll concentration (mg ml^{-1}) is then given by:

$$\frac{(E_{645} \times 20 \cdot 2) + (E_{663} \times 8 \cdot 02)}{20}$$

Demonstration of the ability of the plastids to phosphorylate

When phosphorylation is coupled to photosynthetic electron transport the latter proceeds at a fairly slow rate. The presence of this phosphorylation ability can therefore be demonstrated by adding chemicals which act as uncouplers and measuring the stimulation of electron transport thereby obtained. The following procedure indicates how this should be done and the exercise can be performed if time allows, or in order to demonstrate the uncoupling phenomenon itself.

Electron transport is easily measured by following the rate of reduction of the dye 2,6-dichlorophenolindophenol (DCPIP) upon illumination of the chloroplasts. The uncoupling is accomplished by adding ammonium sulphate.

Set up two cuvettes as below:

	1	2
'Salts' solution	1 ml	1 ml
DCPIP solution	-	0·1 ml
Water	2 ml	1·9 ml

A volume of chloroplast suspension containing about 20 μg chlorophyll should be added to cuvette 1 which is then used as a blank to adjust the spectrophotometer at 600 nm. After addition of a similar quantity of chloroplast suspension, the absorbance of cuvette 2 is measured and remeasured every minute during continuous illumination by a 100-200 Watt bulb placed about 30 cm away.

Repeat the measurement with fresh samples using 0·1 ml $(NH_4)_2SO_4$ and reducing the water by 0·1 ml. The ATP synthesis experiment should not be attempted unless the $(NH_4)_2SO_4$ produces a stimulation of at least 25% in the rate of DCPIP decolorisation. Stimulation of up to 150% may be expected.

AT^{32}P synthesis

Set up a test tube containing the following:

'Salts' solution containing 3 μCi^{32}P	1·0 ml
ADP solution	0·1 ml
Phenazine methosulphate	0·1 ml
Water	3·8 ml

Set up also 5 centrifuge tubes containing 0·1 ml trichloroacetic acid.

Add a convenient volume of chloroplast suspension (equivalent to 50-100 μg of chlorophyll) to the reaction mixture, mix well and withdraw 0·5 ml to centrifuge tube 1. Illuminate the reaction mixture in a rack backed with foil using a 100-250 Watt bulb at a distance of about 12 inches (about 30 cm). Place a glass plate between the lamp and the rack to absorb the heat emitted by the lamp.

Withdraw 0·5 ml aliquots of reaction mixture to centrifuge tubes at 1·5 min intervals until a total of five such samples has been obtained.

Extraction of AT^{32}P

Centrifuge the samples and carefully remove 0·4 ml of the supernatants into numbered stoppered test tubes (minimum volume 15 ml each). Add 1·2 ml acetone, mix and stand for 10 min. Add 2 ml water which has been saturated with isobutanol and hexane. Add 7 ml isobutanol/hexane which has been saturated with water, vigorously mix (using a glass rod or vortex mixer) and allow the phases to separate. Incline the tube and carefully add 0·8 ml molybdate reagent down the side of the tube. Gently mix this with the lower water layer and leave to stand for 5 min. Mix vigorously for 30 sec to extract the phosphomolybdate complex into the upper phase. After separation remove the upper layer into the waste bottle.

Add 0·1 ml of 0·005 M KH$_2$PO$_4$ as carrier and repeat the extraction by adding 7 ml isobutanol/hexane. Mix well, allow the phases to separate and remove the upper layer to waste. The aqueous layer will contain the AT^{32}P, all the inorganic ^{32}P being removed into the upper layers.

Counting and expression of results

Dispense 1 ml aliquots of the aqueous layers into numbered scintillation vials and add 3 ml of 0·02% aqueous methyl umbelliferone. Place in the scintillation counter and determine the activity of each vial. Determine also the background count rate using a blank vial of 1 ml water plus methyl umbelliferone. You will be provided with a standard vial of known isotope content, count this also. Alternatively, the experimental samples and the standard can be counted as dried films on planchets.

Plot counts per minute (after background correction) against time of illumination. Use the results obtained from the standard vial to determine the amount of AT^{32}P synthesised/mg chlorophyll/hour in your system.

Note

It is possible to extend this experiment by including a comparison of the radioactive estimation of ATP synthesis with a chemical method involving a measurement of inorganic phosphate loss by incorporation into ATP.

To do this prepare six tubes containing:

9·5 ml of a solution of 1·58 ml 1% amidol in 20% sodium metabisulphite + 1·58 ml perchloric acid + 100 ml water.

At zero time and at the end of the experiment, 0·3 ml samples of experimental reaction mixture are withdrawn to these tubes; 0·15 ml of 8·3% aqueous ammonium molybdate is added to each and the absorption measured at 660 nm against a water blank.

Student level
HNC to degree students.

Time required
3–4 h

Materials required

Plant material
A variety of plants can be used but since phosphorylation is a delicate process, good quality chloroplasts must be isolated and hence good quality plant material is essential.

Plants should be grown under good conditions and for the experiments young, healthy leaves should be used and the mid-ribs removed after the leaves have been well washed.

Beans, peas, sugar beet, spinach, etc., are suitable, but one of the best materials seems to be commercially available open-heart lettuce. Good quality plants, usually sold in polythene bags, are available in many shops and markets and have the advantage that they quickly show any deterioration in their state.

^{32}P sample
Most samples of ^{32}P contain radioactive pyrophosphate which is not extracted in the solvent extraction process and leads to high levels of activity in the zero time samples.

To remove this, the sample of ^{32}P should be heated in N HCl at 90°C for 60 min. It is then dissolved in an appropriate volume of 'salts' solution and the pH corrected, 30 μCi of ^{32}P dissolved in 10 ml salts solution will provide sufficient material for ten experiments.

It is undoubtedly best if the above procedures are carried out for the students and they are provided with prepared salts solution.

Other requirements

	Per group
100–250 W lamp	1
Clock	1
Glass plate (about 20 cm square)	1
Clamp stand	1
Vortex mixer	1
Plastic centrifuge tubes, 15 ml	5
Glass rods, 6–12 mm diameter	7
Aluminium foil about 30 × 15 cm	
Stoppered test tubes	5
Test tubes in rack	2
Scintillation vials or planchets	6
Spectrophotometer and cuvettes	1
Acetone (80%)	10 ml
Acetone (100%)	7 ml
'Salts' solution	5 ml
4 mM $MgCl_2$ ⎫	
15 mM NaCl ⎪ pH 7·8	
4 mM Na_2HPO_4 ⎬	
50 mM Tricine ⎭	
'Salts' solution + 3 μCi ^{32}P ml^{-1}	1 ml
ADP (10 mM)	0·1 ml
Phenazine methosulphate (0·1 mM)	0·1 ml
Trichloroacetic acid (20%)	1 ml
KH_2PO_4 (5 mM)	1 ml
Water saturated with 1:1 isobutanol/hexane	10 ml
Isobutanol/hexane (1:1)	35 ml
Aqueous methyl umbelliferone (0·02%)	20 ml

Molybdate reagent: 5 g ammonium molybdate +
40 ml 10N H_2SO_4 diluted to 100 ml   5 ml

Optional requirements

(a) Uncoupling demonstration

Cuvettes	2
'Salts' solution	2 ml
2,6-dichlorophenolindophenol (1 mM)	20 ml
Àmmonium sulphate (30 mM)	5 ml

(b) Chemical estimation of phosphate

Test tubes in rack
Molybdate reagent (as described earlier)
Amidol reagent:
1.58 ml of 1% amidol in 20% sodium metabisulphite + 1·58 ml of perchloric acid + 100 ml water. 100 ml.

For general use

Plant material (chilled)
Bottom drive blender (cold)
Ice and ice baths
Refrigerated centrifuge
About 1 m^2 cotton organdie or similar material and 1 m^2 nylon bolting cloth (25 μm mesh size)
Radioactive waste bottle
Pipettes 10 ml, 1 ml, 0·1 ml and automatic
100 ml measuring cylinders
Large filter funnel
250 ml conical flask
Maceration medium (cold)
  0·06 M phosphate buffer pH 7·3
  0·001 M EDTA
  0·5 M sucrose
Resuspension medium (cold)
  0·03 M Tricine pH 7·3
  0·5 M sucrose

THE PHOTOSYNTHETIC UPTAKE OF $^{14}CO_2$ BY LEAVES

Photosynthesis can be divided into two main stages, a light-dependent photoreduction stage, usually measured in experiments involving isolated chloroplasts, and a light-independent carbon dioxide fixation stage, usually followed in whole plants. This experiment involves a very simple procedure to follow the uptake of $^{14}CO_2$ by leaves during photosynthesis and a number of experiments are included to illustrate various aspects of the photosynthetic process in whole leaves.

Basic procedure

The leaves have to be exposed to an atmosphere containing $^{14}CO_2$ under conditions which protect the experimenter but allow the leaf to carry out photosynthesis. The apparatus illustrated in Figure 22 has been found suitable.

It consists essentially of a 'Perspex' '('Lucite') box with a detached lid. A small box (for example, an empty coverslip box) is cemented at one end below a small hole drilled in the

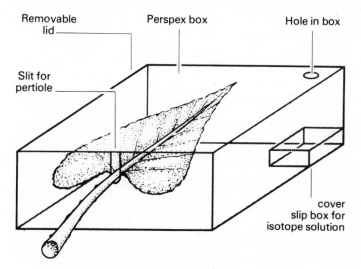

Figure 22. Apparatus for exposing leaves to $^{14}CO_2$.

'Perspex' box. A slot is cut in the lid to take the leaf petiole when the leaves are to remain attached to the parent plant.

The leaf is placed through the lid slot and the lid sealed to the box by 'Parafilm' or 'Sellotape'. The lid slot is sealed similarly, and a small piece of 'Parafilm' placed over the hole above the coverslip box. The box is held horizontally in a clamp and the leaf illuminated evenly by a 60 W bulb placed at a distance of about 30 cm above the box. The leaf can now be exposed to $^{14}CO_2$ and its uptake measured.

Alternatively, if a supply of 'Perspex' boxes is not available, then 250 ml wide necked conical flasks will suffice. The petiole can be passed through a longitudinally split, bored rubber bung and sealed well with 'Vaseline' or 'Parafilm'. The two liquids can be introduced through a hypodermic syringe needle placed permanently in the bung and if the flask is held at a 45° angle these will run to the bottom of the flask and not come into contact with the leaf.

EXPERIMENT 1.   PHOTOSYNTHESIS BY A GREEN LEAF

Detach a medium-sized healthy leaf from a suitable plant (*Geranium*, green ivy, etc.) and place in the apparatus as described above, trying to ensure that the undersurface of the leaf is raised from the 'Perspex' box. Seal the box well and switch on the fume cupboard pump. Cover the end of the cut petiole with 'Vaseline' to help prevent the leaf drying out.

Using a hypodermic syringe take up a volume of $NaH^{14}CO_3$ solution containing about 10 μCi of activity. Inject it through the 'Parafilm'-sealed hole into the small box. Follow this by injecting a small volume (about half the volume of $NaH^{14}CO_3$ solution taken) of concentrated HCl into the small box. This will release the $^{14}CO_2$ from the bicarbonate but the amount of HCl fume is negligible and does not affect the leaf. Quickly seal the needle perforation with a small piece of film or 'Vaseline'. Illuminate the leaf evenly for about 15 min and after this time switch off the lamp, open the box, and remove the leaf. Place the leaf on a porcelain tile and remove the residual $NaH^{14}CO_3$ solution from the box or flask to a waste bottle using a Pasteur pipette.

If a sensitive monitor is available it can be used to show

that all of the $^{14}CO_2$ has been taken up by the leaf and that the latter is now radioactive. This illustrates quite well the considerable CO_2 scavenging potential of actively photosynthesising leaves. It also shows that little if any $^{14}CO_2$ enters the atmosphere, and by carrying out the experiment in a fume cupboard any that does, should not contaminate the laboratory atmosphere.

Using a small cork borer (e.g. No. 6) cut discs of about $\frac{1}{2}$-1 cm diameter out of the leaf. It is desirable to cut at least two discs from roughly the same position on either side of the mid-rib, as replicates. Place the discs on dimple planchets and count each of them twice for 100 sec on each occasion. Average the duplicate counts and subtract the background count. Care should be taken in carrying and handling the planchets within the counting room because the low energy of the ^{14}C radiation makes it impossible to carry out the recommended practice for loose samples, which is to cover them with 'Sellotape'. The discs may also be counted by placing at the bottom of small diameter tubes ('liners') inserted into a normal scintillation vial.

EXPERIMENT 2. REQUIREMENT FOR LIGHT

The requirement for light can easily be shown by covering half the top surface of a leaf tightly but gently with aluminium foil and arranging the leaf in the box so that the foil faces the light. The restriction on photosynthetic gas flow is minimal since the majority of stomata are on the lower surface of the leaf. Light reflections from the box or flask can be reduced by using black paint or paper.

After exposure to $^{14}CO_2$, two discs are cut from the light side and two from the dark side of the leaf and counted as before. If the leaf is dried between filter paper and glass plates in an oven at $100°C$ for 30 min in order to kill it and fix the ^{14}C in place it may be subjected to autoradiography as detailed earlier (Chapter 7). After one or two days exposure, the difference between the exposed and unexposed area will be quite apparent.

EXPERIMENT 3.   PHOTOSYNTHESIS IN VARIEGATED LEAVES

(a) The requirement for chlorophyll can be shown by using variegated leaves of the yellow/green kind in this experimental system. Variegated ivy has proved satisfactory and is available during the winter.

The experiment is carried out exactly as described above. After the removal of the leaf from the box a sensitive end window counter placed against the different leaf regions will usually detect the different isotope levels.

Cut at least two discs from each region, trying to take them from areas opposite to one another on either side of the mid-rib. After removing the discs count them as above and then prepare an autoradiogram of the leaf. One or two days exposure will be sufficient to show intense activity in the green area and the activity moving along the veins into the much less active yellow areas.

(b) If a red/green *Coleus* or *Begonia* leaf is used and the uptake measured as in experiment (a), it is usually found that while the uptake in the red area is less than in the green portions (perhaps 50%) less) it is considerably more than in the yellow area of yellow/green variegated leaves (usually 5–10% of the green area activity).

This indicates the presence of chlorophyll and active photosynthesis in the red areas, the rate being lower due to a lower level of chlorophyll or masking of light absorption by the red pigments.

EXPERIMENT 4.   TRANSPORT OF PHOTOSYNTHATES WITHIN THE LEAF

Experiments concerned with transport of photosynthetic materials within plants are not easily performed. However, the availability of variegated leaves enables a very simple demonstration of the movement of photosynthates to be made.

Two variegated ivy leaves should be taken and set up in the normal experimental system. After 15 min of photosynthesis in $^{14}CO_2$ one leaf is removed and two discs are cut from each of the yellow and green areas, if this is possible, and their activities determined. The rest of the leaf is then autoradiographed. The other leaf is left, still sealed in its box, under illumination for 12–24 h. During this time, transport of photosynthates from green to yellow areas occurs, and since the leaf is detached its activity is not declining due to transport of materials into the stem. Much of the $^{14}CO_2$ evolved by the leaf during its respiration will be taken in again during photosynthesis since the leaf is being constantly illuminated in a closed environment. After the 12–24 h period, this second leaf is treated in the same way as the first.

An examination of the leaf activities shows that the difference between yellow and green areas is very much less in the second leaf. The autoradiogram also shows this difference and in addition it becomes apparent that much of the yellow area activity is concentrated in the leaf veins and rather less in the cells themselves.

EXPERIMENT 5.   EXPERIMENTS ON STOMATA
Photosynthesis is carried out in chloroplasts, packed into the cells within the leaves. The required gas exchange between the chloroplasts and the environment occurs through special epidermal perforations termed stomata. These are found principally on the lower surface of the leaves and are essential to prevent excessive water loss by the leaves.

It is easy to carry out a number of experiments to show the importance of the stomata and some factors controlling their closure.

(a) Importance of the stomata. Stomatal importance can easily be demonstrated by artificially closing or blocking them. At the same time their distribution on the leaf can be shown.

Two Geranium leaves should be selected, and on one, half of the lower surface of the leaf should be covered with a thin film of 'Vaseline'. The other should have half of the upper surface similarly covered. They are then placed in separate boxes and their photosynthesis measured in the previously described experimental system, either by autoradiography or by counting the activity of leaf discs. (During autoradiography place the ungreased surfaces next to the film.)

Only when the lower surface is greased is photosynthesis considerably inhibited, indicating that the greatest concentration of stomata occurs here. Interestingly, diffusion of CO_2 taken in through the open stomata to the greased region via the intercellular spaces of the leaf mesophyll is low, but slight diffusion is often apparent on autoradiograms.

(b) *Low relative humidity*. Very dry air will of course tend to dessicate most leaves fairly quickly, and when the plant reaches a low water tension the stomata close in order to conserve water supplies. The uptake of CO_2 and photosynthesis is, unfortunately for the plant, inhibited at the same time.

Two *Geranium* leaves of similar size and age are detached from a single plant and sealed into separate boxes, one of which contains a layer of activated silica gel granules. After 2 h the experiment is carried out as before. The conditions employed do not usually bring about complete stomatal closure but will reduce $^{14}CO_2$ uptake in the dessicated leaf. The different levels of photosynthetic activity can be detected by either counting pairs of cut discs or by autoradiography.

It is interesting to try the same experiment using an ivy leaf. The xerophytic nature of the leaf makes it much more resistant to dessication.

(c) *Closure by chemical agents*. Some chemicals affect the processes involved in opening and closing the stomata and in fact on occasion these have been used as aerial sprays to reduce transpirational water loss in arid areas. Phenylmercuric acetate is one of these chemicals and may act by interfering with guard cell photosynthesis.

A single *Geranium* leaf is selected, allowed to remain attached to the plant and divided by an ink marking pen into three equal parts. These should run radially from the point of petiole attachment. The upper surfaces should be marked 1, 2 and 3 using small numbers. Zone 1 should have both its surfaces painted with a 1% solution of detergent, zone 2 with 5×10^{-5}M phenylmercuric acetate in 1% detergent and zone · 3 left untouched.

The plant bearing the treated leaf should be left overnight

for the treatments to take effect. The leaf can then be detached, inserted into an experimental box, sealed, and its photosynthetic rate measured as usual.

After the experiment the leaf should be well washed in a beaker of water and two or three discs should be cut from each zone, avoiding areas marked with the pen. On determining their activities it becomes evident that treatment with phenylmercuric acetate does result in a decrease in photosynthetic rate in that area of the leaf.

It is well to bear in mind that phenylmercuric acetate is poisonous and should not come into contact with the skin.

This experiment does not, of course, distinguish between direct effects on stomata and the known inhibitory effects of phenylmercuric acetate on photosynthesis. It seems, however, to be very likely that phenylmercuric acetate does not penetrate beyond the epidermal cells to the leaf mesophyll.

Student level
Six-formers and above.

Time required

Experiments 1, 2, 3, 5a	45 min.
Experiment 4	45 min. Autoradiography overnight, 1 h for film development
Experiment 5b	Preparation 2 h beforehand, then 45 min.
Experiment 5c	Treatment overnight, then 45 min.

Materials required

	Per group		
'Perspex' box prepared as described, or a 250 ml conical flask and split bung	1	Sharp number 6 cork borer	1
Clamp stand	1	Forceps	1
60 W lamp	1	Rubber gloves	1 pair
Stop clock	1	Small paint brush	1
White tile	1	$NaH^{14}CO_3$ solution of known activity (per experiment)	$10\,\mu Ci$

Concentrated HCl in small flask	5 ml
Detergent, 1%	5 ml
Phenylmercuric acetate (5 × 10⁻⁵ M) in 1% detergent	2 ml
Activated silica gel granules	50 g
1 ml hypodermic syringes	2

For general use

A supply of suitable plants. These should bear a number of healthy, medium-sized leaves and be well watered.

'Parafilm' or 'Sellotape'

'Vaseline'

Fume cupboard

Pasteur pipettes and teats

Radioactive waste bottles

Dimple planchets, or scintillation vials, liners and scintillant

Autoradiography film (10 cm square), preparation, handling and developing facilities (chapter 7)

Aluminium foil

Glass plates (10 × 10 cm)

Filter papers (15 cm)

Oven at 100°C.

AN INVESTIGATION OF THE CALVIN CYCLE USING $^{14}CO_2$

The experiment described illustrates the method used by Calvin and his co-workers when they elucidated the CO_2 fixation cycle of photosynthesis. In using biochemical techniques involving the culture and manipulation of a unicellular alga the experiment is also useful in that it links the two fields of biochemistry and cell biology for the student.

For some time it has been known that photosynthesis consists basically of two stages:

(*a*) a light-dependent one in which light is used to generate ATP and reducing power in the form of NADPH

(*b*) a light-independent one in which these components are used to convert CO_2 to a number of photosynthetic products.

Calvin and co-workers realised that after exposing a photosynthetic organism (a unicellular alga for convenience) to $^{14}CO_2$ for various periods of time then the ^{14}C in each case is incorporated to different 'distances' along the metabolic route concerned. When photosynthesis is stopped by immersing the alga in hot alcohol the labelling pattern is 'frozen'. The earlier a compound appears in the metabolic pathways the more intense its labelling should be at any one time when compared with compounds later in the sequence (assuming that the pool sizes are roughly similar). Also as the period of $^{14}CO_2$ exposure lengthens additional labelled compounds appear as the ^{14}C moves down the metabolic pathways.

A major problem was to separate, identify and quantitatively determine the potentially large number of possible products of photosynthesis. The separation was achieved by two dimensional paper chromatography and the components located and identified using various spray reagents, by comparison of R_f values with published figures and by elution followed by various chemical tests and co-chromatography with authentic compounds.

The radioactive spots were visualised by autoradiography and the developed X-ray film superimposed on the chromatography paper to locate the spots. The intensity of the radioactivity could be gauged by eye from the autoradiogram or more quantitatively by eluting the spots from the paper and counting them in a scintillation counter.

The experiment described here illustrates fairly simply these procedures.

Procedure

1. *Preparation of the algal culture*
A suitable alga is *Chlorella pyrenoidosa* and it can be grown on Knops solution in a warm, light place such as a window sill, although more dense cultures can be obtained and with greater speed by using an illuminated orbital shaker. Before the laboratory period, the cells are centrifuged off at low speed and resuspended in fumarate buffer. This removes the salts present in the original medium since they may interfere with chromatography at a later stage.

It is important not to use too many algal cells if the chromatography is to be successful; 20 mg has proved to be a suitable quantity. In order to determine the volume of algal suspension equivalent to this, a suitable volume of suspension is dried in an evaporating basin and weighed.

2. *Treatment with $^{14}CO_2$*
Assemble the apparatus shown in Figure 23 and ensure that all the joints are well made. If 'Quickfit' apparatus is

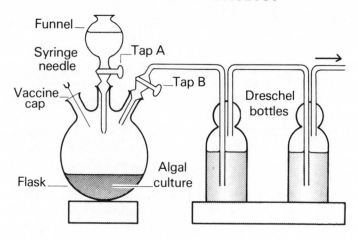

Figure 23. Apparatus for demonstrating the Calvin cycle.

not available make use of good quality, well fitting rubber bungs and seal all joints with 'Vaseline'.

Dilute the 20 mg of algal cells to a suitable volume (25–50 ml) with fumarate buffer and add this to the 3-neck flask and stir vigorously. Connect up the rest of the apparatus, close taps A and B and leave the pump switched off. Illuminate the flask with a 60 W bulb placed 25 cm from it for 5 min before adding the $NaH^{14}CO_3$.

Carry out the following three operations:

(*a*) Check the apparatus over and make sure it is not likely to come apart.

(*b*) You will be provided with a solution of radioactive sodium bicarbonate ($NaH^{14}CO_3$; 100 μCi). Take up the solution in a hypodermic syringe, carefully remove the needle and connect the syringe to the needle through the vaccine cap in the apparatus. Leave it there for the present. Ensure that the solution does not contact your hands and if any is spilt use a tissue to mop it up and dispose of it as radioactive waste.

(*c*) Have boiling on a hot plate a quantity of methanol equal to the volume of *Chlorella* cells to be used (normally 25–50 ml).

Inject the radioactive bicarbonate solution into the photo-synthesising *Chlorella* cells and start the clock.

Carefully place the hot alcohol into the separating funnel and after a suitable period of time (0· 5-5 min) open taps A and B, in that order. Switch on the vacuum pump and switch off the light. The alcohol will stop further photosynthesis by killing the algae and will extract the cell contents. Draw air through the apparatus for a minimum of 15 min while main-taining vigorous stirring of the culture. Residual $^{14}CO_2$ will be drawn off and safely trapped as barium (or sodium) carbonate.

Detach the flask from the rest of the apparatus. Place the latter to one side bearing in mind that it will be con-taminated. Centrifuge the contents of the flask at 1,000 g for 5 min and remove the supernatant. Evaporate the super-natant on a hot plate using initially a large beaker followed by an evaporating basin. Continue until a volume of 2-3 ml is obtained, *do not* allow to dry out and be careful to avoid the liquid spitting in the final stages.

3. *Paper chromatography*

Take a piece of Whatman No. 1 chromatography paper of a suitable size (20 or 30 cm square) and 5 cm from the bottom and 5 cm from one side mark a cross lightly with a pencil. Write on the paper your name, date and direction of development of the two solvents.

On the pencil cross spot 10 μl of the extract keeping the spot as small as possible. Develop the paper in the phenol solvent; dry it in a forced draft for as long as possible (10-18 h) and then develop the paper again in butanol/acetic acid/water solvent at right angles to the first solvent.

In each case develop the paper until the solvent is near the top and be careful to dry the paper thoroughly after the second development.

Prepare an autoradiogram of this paper using Kodirex X-ray film as directed earlier (Chapter 7). Leave for 1 month, then develop and fix it as described.

The origin of the chromatogram should be visible as an intense spot and several other spots should also be visible. If a range of incubation times has been used by different groups of students a difference in complexity of the various chromatograms should be apparent.

Suggestions for further work

Attempts can be made to identify the substances present on the chromatogram by various spray reagents. Details of these can be found in the standard texts on paper chromatography. It has to be borne in mind that autoradiography is a very sensitive technique and insufficient material may be present in the chromatogram to produce a visible spot with the spray. The R_f values on the paper and autoradiogram may be compared with standards or published data.

Student level
HNC to degree students

Time required
The experiment and preparation of extracts 3–4 h
Chromatogram development 2–3 d
Autoradiogram exposure 3–4 wk
Autoradiogram development 1 h

Materials required

	Per group		
250 ml 3-necked flask	1	Absolute methanol	50 ml
250 ml separating funnel	1	Barium (or sodium)	
250 ml Dreschel bottles	2	hydroxide	500 ml
250 ml beaker	1	Buffer, 1·8 mg sodium	
50 ml measuring cylinder	1	hydroxide and 5 mg fum-	
1 ml hypodermic syringe		aric acid in 10 ml water	100 ml
(and two needles)	1	NaH^{14}CO$_3$	100 μCi
Vaccine cap	1	Whatman No. 1	
Cone/tube bend with tap	1	chromatography paper 20	
Vacuum pump	1	cm × 20 cm	1
Lamp, 60 W	1	Micropipette, 10 μl	1
Stirrer and follower	1	X-ray film	1 sheet
Tray	1	Glass plates	2
30 cm pressure tubing	2 lengths	Storage box	1
Clock	1	*For general use*	
Evaporating basin	1	Fume cupboards	
Hot plate	1	Algal cultures of known dry	

weight/ml
Bench centrifuges to take 100 ml
tubes
Chromatography tanks for 2-D
ascending chromatography
Solvents (a) Butanol (4)/Acetic
acid (1)/Water (5)

(b) Phenol (225 g)/water (25 ml)/
Ammonia (2·5 ml)
Dark room, safe light and develop-
ing facilities
'Sellotape' and scissors
Pipettes, graduated, various sizes

THE ABSORPTION OF ^{32}P BY YEAST

The two possible mechanisms by which yeast could absorb anions from the environment are passively, in response to electrical or osmotic gradients, or actively, involving a definite energy-requiring mechanism within the membrane. In the following experiments, the uptake of $^{32}PO_4^{3-}$ by yeast cells growing under different conditions is examined mainly in order to demonstrate the involvement of cellular energy in the process and by inference, an active transport system for the ion.

Active transport mechanisms, as the name implies, require considerable amounts of energy in the form of ATP. Yeast cells can supply the ATP required for active transport *either* (a) by aerobic respiration, a process involving electron transport linked phosphorylation (and indirectly the operation of the Krebs cycle) *or* (b) anaerobically by an ethanol-yielding fermentation wherein ATP is generated by substrate level phosphorylation. In one experiment the effect of a glycolysis (and fermentation) inhibitor (KF) and a Krebs cycle inhibitor (KCN) are investigated.

The production of energy in the form of ATP is a metabolic process and as such is strongly influenced by the environmental temperature. An experiment is described which will readily show that as the temperature falls the rate of uptake of ^{32}P also falls; strongly implicating cellular energy and enzymes as vital components in anion uptake.

A ready source of energy for yeast cells is glucose and another experiment is included which measures the uptake of ^{32}P with various concentrations of added glucose.

If an active transport mechanism is involved in phosphate

uptake, then the phosphate ions probably combine with a suitable membrane-based carrier, the complex then crossing the membrane before releasing the phosphate into the cell. Such carriers are not always absolutely specific for a given ion and the presence of such a carrier can be inferred from a demonstration of the reduction in phosphate uptake (because of competition for the carrier) in the presence of the similar ion arsenate.

Yeast cells are ideally suited to these experiments because, apart from being cheap and easy to grow, they are tough and capable of withstanding severe environmental changes. Their very strong cell wall for instance enables them to withstand large osmotic pressure changes. They also have low passive permeability to phosphate.

These experiments illustrate quite well a considerable advantage of using isotopes since it is necessary to measure the movement across the cell membrane of small amounts of phosphate, when phosphate already exists on both sides of the membrane. Such experiments are really only possible with radioisotopes since only they can label the ion without changing its physical or chemical properties.

EXPERIMENT 1.　THE EFFECT OF RESPIRATION INHIBITORS

Set up eight, numbered, 100 ml conical flasks with the following contents (ml):

	1	2	3	4	5	6	7	8
Glucose	0	10	10	10	10	10	10	10
Water	12	2	1·9	1·9	1	1	1	0
NaH$_2$PO$_4$ (0·05 M)	3	3	3	3	3	3	3	3
NaH$_2$PO$_4$ (0·05 M + 1 μCi^{32}P ml^{-1}	3	3	3	3	3	3	3	3
Arsenate (0·3 M)			0·1					
Arsenate (3 M)				0·1	1			
KCN						1		1
KF							1	1
Yeast	2	2	2	2	2	2	2	2

Add the contents in the order given, mixing the flask

contents before adding the yeast suspension. Place in a 30°C shaking water bath and leave for 1 h.

Remove 2 ml of the yeast suspension to numbered centrifuge tubes containing 5 ml of unlabelled NaH_2PO_4. Centrifuge at full speed on a bench centrifuge for 5 min. Carefully remove the supernatant into a container for radioactive liquid waste. Add a further 5 ml of NaH_2PO_4 to each tube and, after covering the top of the tube with water-proof tape, use a vortex mixer to mix the contents. Recentrifuge and remove the supernatant to waste as before. Add 0·5 ml dilute detergent to each tube and transfer the contents to numbered planchets using separate Pasteur pipettes. Carefully evaporate the samples to dryness under a heat lamp. Rinse out the tube with a further 0·5 ml detergent and evaporate on to the planchets as before.

Use an end window counter to count each planchet twice for 100 sec on each occasion. Average each of the duplicate counts and subtract the average value for flask 1 from those for flasks 2–8. Flask 1 provides an indication not only of the background, but also of the activity due to the phosphate ions trapped between, and adsorbed on to, the yeast cells and absorbed by the cell in the absence of energy supplying substrate. Any increase above this value must be due to uptake in the presence of glucose and the other materials.

If scintillation counting is to be used, 1 ml water can be added to each of the tubes containing the washed yeast cells, in the place of detergent, and the contents transferred directly to numbered scintillation vials. Methyl umbelliferone solution is added and they are counted directly using the Cerenkov technique. Settling of the cells does not affect the counting efficiency.

EXPERIMENT 2.   THE EFFECT OF VARIATION IN GLUCOSE
CONCENTRATION

Label seven, 100 ml conical flasks 1-7 and pipette in the
following solutions (ml):

	1	2	3	4	5	6	7
0·05 M NaH_2PO_4 + $^{32}PO_4$ (1μCi ml—^{1})	3	3	3	3	3	3	3
0·05 M NaH_2PO_4 (Unlabelled)	3	3	3	3	3	3	3
Glucose	0	0·2	0·4	0·6	0·8	1·0	1·2
Water	12	11·8	11·6	11·4	11·2	11·0	10·8
Yeast suspension	2	2	2	2	2	2	2
FINAL GLUCOSE CONCENTRATION (mM)	0	5	10	15	20	25	30

After adding the yeast suspension, incubate the flasks in a
shaking water bath at 30°C for a suitable time interval up to
1 hour. Analyse the samples in the manner described in
Experiment 1.

EXPERIMENT 3.   THE EFFECT OF VARIATION IN TEMPERATURE

Set up six, numbered 100 ml flasks each containing the
following:

Glucose	10 ml
NaH_2PO_4 (unlabelled)	3 ml
NaH_2PO_4 + $^{32}PO_4$ (1 μCi ml^{-1})	3 ml
Water	2 ml

Put one flask into each of the following environments, 0°C
(ice/water), room temperature, 30°, 40°, 50° and 60°C.
Pipette 4 ml 20% yeast into each of six test tubes and place
one into each of the above environments. Leave all the flasks
and tubes for 10 min to equilibrate, then pipette 2 ml of
appropriate yeast suspension into the corresponding flasks.
Leave all the flasks for 1 h with occasional shaking.

Analyse the samples in the manner described in
Experiment 1. Proper controls for this experiment would
make it cumbersome and useful results can be obtained by
using the experimental results directly.

Student level
HNC and degree level.

Time required

Experiment 1	$2\frac{1}{2}$h
Experiment 2	3 h
Experiment 3	$2\frac{1}{2}$–3h

Materials required

Experiment 1

Per group

Glucose (0·5 M)	5 ml	*or* scintillation vials,	
NaH$_2$PO$_4$ (0·05 M)	25 ml	liners and 0·02% methyl	
NaH$_2$PO$_4$ (0·05 M) con-taining 1 μCi/ml^{-1} ^{32}P	25 ml	umbelliferone	8
		Water bath at 30°C	1
Sodium arsenate (3 M)	1·1 ml	*For general use*	
Sodium arsenate (0·3 M)	0·1 ml	Bench centrifuges	
Potassium cyanide (0·6 M)	2ml	Isotope warning tape	
Potassium fluoride (0·7 M)	2ml	Vortex mixers	
Yeast suspension (20%)	16 ml	Pipettes—graduated 10 ml, 5 ml	
100 ml conical flasks	8	0·1 ml	
10 ml centrifuge tubes	8	Dilute detergent	
Planchets and heat lamps		Pasteur pipettes and teats	

Experiment 2

Per group

Glucose (0·5 M)	42 ml	umbelliferone	7
NaH$_2$PO$_4$ (0·05 M)	140 ml	Shaking water baths at 30°C	1
NaH$_2$PO$_4$ (0·05 M) con-taining 1 μCi ml^{-1} ^{32}P	21 ml	*For general use*	
Yeast suspension	14 ml	Pasteur pipettes and teats	
100 ml conical flasks	7	Bench centrifuges	
10 ml centrifuge tubes	7	Vortex mixers	
Planchets and heat lamps		Isotope warning tape	
or scintillation vials, liners and 0·02% methyl		Bottles for radioactive waste	
		Dilute detergent for planchets	

Experiment 3

	Per group	
Glucose (0·5 M)	60 ml	umbelliferone 6
NaH$_2$PO$_4$ (0·05 M)	20 ml	*For general use*
NaH$_2$PO$_4$ (0·05M) containing 1 μCi ml^{-1} ^{32}P	18 ml	Water baths at 30°, 40°, 50° and 60°C
Yeast suspension	12 ml	Ice
100 ml conical flasks	6	Bench centrifuges
Test tubes	6	Isotope warning tape
10 ml centrifuge tubes	6	Vortex mixers
Planchets and heat lamps *or* scintillation vials, liners and 0·02% methyl		Waste bottles Dilute detergent Pasteur pipettes and teats

Yeast suspension

Dried yeast is suitable but it must be allowed to metabolise, in aerated suspension, for at least 18 h to remove yeast storage products and the sugar in which the dried yeast is packed.

THE UPTAKE AND TRANSLOCATION OF ^{32}P BY SUNFLOWER SEEDLINGS

Growing plants require phosphorus for a number of important syntheses including the synthesis of ATP for energy transfer, nucleic acids such as DNA and phospholipids for membranes.

Plants take up phosphorus mainly in the form of phosphate from the soil and it is then translocated via the xylem system predominantly to the growing points (meristems) and points of active synthesis (e.g. young leaves). Thus, by immersing the roots of sunflower seedlings in a solution of phosphate labelled with ^{32}P for a short time and then placing them into unlabelled phosphate for varying lengths of time, it is possible to follow the uptake and movement of the ^{32}P.

Basic procedure

You will be provided with a 25 ml conical flask containing

20 ml 0·05 M NaH_2PO_4 pH 7 and ^{32}P; and also with a pot of sunflower seedlings.

Lay the pot on its side and ease out its contents on to a large paper towel. Select five seedlings of about the same size and while avoiding damage to the root system wash the compost off the roots in a large beaker of water.

Place the seedlings in a small beaker of non-radioactive phosphate solution and illuminate with a 60 W bulb at a distance of about 30 cm for 5-10 min to allow the plants to settle down. During this time, place four 500 ml beakers of water on paper towels and touching one another.

At zero time, place one seedling into the radioactive phosphate, illuminate it as before, and leave for exactly 2 minutes.

Remove the seedling and rinse the roots quickly in each of the beakers of water in turn to remove the excess radioactive phosphate. Avoid letting the liquid drop on to the bench.

Place the seedling between sheets of filter paper, then between two glass plates (label with the time) and place a heavy weight on top of the plates. Heat in an oven at 100°C for 15-30 min until the plants are dry.

Place a second seedling into the radioactive phosphate, illuminate and leave for 2 min. Remove the seedling, wash as before then place it into a flask containing non-radioactive phosphate for a further 2 min. Heat press as before.

Repeat this procedure with five other seedlings leaving them in the non-radioactive phosphate for 8, 28, 58 min and if possible overnight. Remove each one, wash and heat press as before.

Autoradiography

After drying the seedlings, carefully separate the sheets of filter paper. The seedlings will probably have stuck to them and will break if removal is attempted; therefore, carefully cut round each seedling with a pair of scissors leaving as little as possible of the filter paper. Label each seedling clearly with the time period for which it was immersed in the phosphate buffer. Keep the seedlings between glass plates until they are all dried.

Using forceps NOT fingers arrange the six dry seedlings on a 25 × 20 cm piece of paper, labelling each clearly, and attach them with small pieces of 'Sellotape' or small amounts of glue.

^{32}P emits β-particles of high energy and therefore close contact with the dried plants should be avoided.

In a dark room, using a Kodak 6B safelight, place a 25 × 20 cm 'Kodirex' X-ray plate carefully over the dried seedlings. Sandwich this between two plates of glass (25 × 20 cm) and tape them together. Place this sandwich in a light-proof box for 7 days.

Develop, fix and dry the plate (Chapter 7) after first cutting a small piece off the top right-hand corner. Re-locate the plate over the seedlings with the cut corner in the correct position. With a felt-tip pen draw round the outline of the seedlings on the X-ray plate. Label each drawing with the time of experimental exposure.

Radioactivity determination

A quantitative indication of radioactivity can be obtained by cutting discs (about 0·5 cm) with a cork borer and counting them. It is best if a disc is cut from each of two opposing leaves and counted either on a dimple planchet covered with 'Sellotape' or in a scintillation vial with a liner using the Cerenkov technique.

EXPERIMENT 1.   THE EFFECT OF TRANSPIRATION RATE ON ^{32}P UPTAKE

Since the phosphate is transported via the xylem system any factor influencing transpiration will affect the rate of trans-location of the ^{32}P. This experiment therefore offers an alternative to instruments such as the potometer in studies of transpiration, and the effect of conditions such as light, dark, humidity, stomatal closure, etc., can easily be studied.

Procedure

Nine to ten seedlings are prepared in the manner described in the basic procedure. One of these seedlings, in a beaker of buffer, is placed in the dark for at least an hour before using it.

You will be provided with the apparatus shown in Figure 24; carefully place a seedling in it, as shown, having first rinsed the boiling tube in hot water to increase the humidity. The flask should contain 20 ml phosphate buffer. Leave the apparatus as long as possible (at least 30 min) before use, then add 10 μc ^{32}P in 1 ml buffer to the flask and leave for a further 15 min. Quickly remove the seedling, wash and press as in the previous experiment; carefully labelling the sandwich with the treatment given to the plant.

Pour the labelled buffer into a 50 ml conical flask and keep it for the procedures listed below.

Take six more seedlings from the beaker of unlabelled phosphate buffer and place them in 50 ml conical flasks containing 20 ml phosphate buffer (unlabelled) and treat them so:

1. Grease the upper surfaces of the leaves.
2. Grease the lower surfaces of the leaves.
3. Grease both surfaces of the leaves.
4. 'Ring' the stem with melted paraffin wax being careful not to overheat the wax otherwise the stem will break.
5. Leave on the bench, out of draughts and in the light.
6. Blow air over the leaves from a hair dryer.

Leave the seedlings under the conditions specified for at least 30 min to equilibrate before using, then remove each in turn to the flask containing the ^{32}P and leave in this for 15 min while keeping the seedlings under the conditions specified during this time. Finally, treat the seedling that has been kept in the dark in a similar manner although maintaining it in the dark.

After treatment remove each seedling, wash the roots and cut a disc from each leaf using a suitably sized cork borer. Use forceps to place the discs on labelled planchets, cover them with 'Sellotape' and count the discs twice for 100 sec on each occasion. If the plant is to be autoradiographed, the remainder of each plant should be heat pressed and treated in the manner described in the basic procedure.

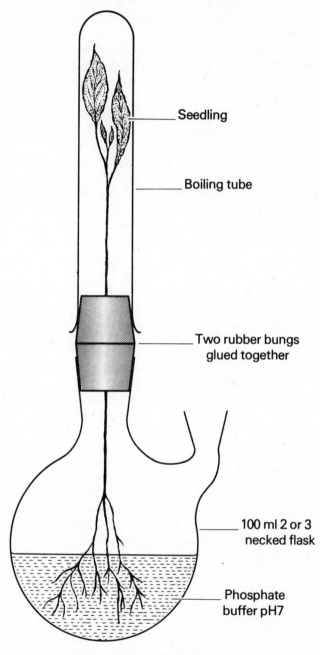

Seedling

Boiling tube

Two rubber bungs
glued together

100 ml 2 or 3
necked flask

Phosphate
buffer pH7

Figure 24. Apparatus for keeping seedlings in 100% humidity.

EXPERIMENT 2. DETERMINATION OF THE WEIGHT OF ^{32}P
TAKEN UP BY THE SEEDLINGS

If an experiment involving the uptake of ^{32}P labelled phosphate by seedlings is performed it is very instructive to determine the quantity of radioactive phosphate in the leaves at the end of the experiment since this illustrates the extreme sensitivity of the techniques of autoradiography and Geiger-Muller or scintillation counting.

Procedure

After drying the labelled plants between filter papers, and carrying out autoradiography and/or disc assay; cut off the remains of two opposite leaves from one of the plants, handling them with forceps and *not* with fingers. Place the leaves in separate 10 ml conical flasks and if the plants have been, or for other reasons are, stuck to filter paper, this can be added also.

Add 2 ml concentrated nitric acid to each flask and heat on a hot plate in a fume cupboard. Avoid the acid boiling vigorously and remove the flask with tongs after the leaf (and paper if present) has disintegrated. This should only take a few minutes.

Using a Pasteur pipette, transfer the contents of the flask to a small measuring cylinder and make up to a convenient volume (say 2·5 or 3 ml) with concentrated HNO_3; this should preferably be acid that has been used to rinse out the conical flask first.

Dry suitable volumes of the liquid on to aluminium planchets, count the samples twice over a fairly long time interval (say 300 sec), average and subtract the background count rate. Alternatively, Cerenkov counting in a liquid scintillation counter can be used.

A sample of known activity, either dried on to a planchet or placed in a scintillation vial, will be supplied. Count this sample in the same way as your experimental samples.

Calculation

In order to be able to calculate the quantity of ^{32}P-phosphate taken up it is necessary to know the specific activity of the sample obtained from the radiochemical suppliers. Details of this are not always provided with the sample but the suppliers will give it if requested or 50 Ci mg—^{1} can be taken as a reasonable value.

It is also necessary to know the efficiency of the instrument used for detection of ^{32}P. This is determined from the count rate obtained for the standard sample.

Thus, if the standard sample contains $0 \cdot 1$ μCi of activity, this is equivalent to $3 \cdot 7 \times 10^3$ disintegrations per second. The instrument efficiency (Y) for ^{32}P is given by

$$\frac{\text{counts per second for standard}}{3 \cdot 7 \times 10^3} \times 100$$

If x = counts per second detected for an *entire* experimental sample, then X', the disintegrations *actually* emitted by the sample is given by:

$$X = \frac{100}{Y} \bullet x$$

where Y is the percentage efficiency of the detector system.

Now 1 Curie produces $3 \cdot 7 \times 10^{10}$ disintegrations per second and therefore the sample contains

$$\frac{X'}{3 \cdot 7 \times 10^{10}} \text{ Curies of activity i.e. } \frac{X'}{37} \text{ nCi of activity}$$

If the specific activity is given as being Z Ci mg^{-1} phosphate, this is

$$Z \times 10^9 \text{ n Ci mg}^{-1} \text{ or } \text{ ZnCi pg}^{-1}$$

then the sample contains

$$\frac{X'}{37 \times Z} \text{ pg }^{32}\text{ P}$$

It is quite astounding to find that the quantities obtained are usually in the picogram range. The methods used in this procedure are perhaps fairly crude, and the losses and assumptions made fairly noticeable. However, even if the results are several hundred percent in error the quantities obtained are still exceedingly low and illustrate the value of using isotopes in this type of work.

Using duplicate leaves gives an indication if errors are likely to be significant.

Student level
Sixth-formers to degree students

Time required
Basic experiment 3-4 h.
Transpiration experiments 3-4 h.
^{32}P weight determination 1-2 h.
Optional autoradiography: preparation 1½h.
  exposure 7 d.
  development 1½h.

Materials required

	Per group		
Basic experiment			
Sunflower seedlings about 1-2 weeks old grown in peat or vermiculite	8-10	Sunflower seedlings about 1-2 weeks old grown on peat or vermiculite	8-10
25 ml conical flask containing 0·05 M NaH$_2$PO$_4$ brought to pH 7 with NaOH. 10 µCi ^{32}P added	1	NaH$_2$PO$_4$ pH 7, 0·05 M (unlabelled)	200 ml
NaH$_2$PO$_4$, 0·05 M (un-labelled)	200 ml	NaH$_2$PO$_4$ pH 7, 0·05 M containing 10 µCi ^{32}P	20 ml
		50 ml conical flasks	7
60 W lamp	1	500 ml beaker	1
100 ml conical flasks	5	100 ml beakers	2
100 ml measuring cylinder	1	Hair dryer	1
Glass chromatography plates (20 × 20 cm)	12	Glass plates (20 × 20 cm)	12
		Heavy weight	1
Heavy weight	1	Forceps	1
Forceps	1	Clock	1
Clock	1	Light-proof box	1
		Weight determination	
Transpiration experiment		Scissors	1
Apparatus shown in the diagram. The bungs are joined with contact ad-hesive, cut longitudinally and grooved to take plant		Forceps	1
		10 ml conical flasks	2
		Hot plate and crucible tongs	1
	1	^{32}P standard	1

Pasteur pipettes	2	Concentrated HNO_3	10 ml
5 ml measuring cylinders	2		

General requirements for all experiments
Large paper towels
Ovens at 100°C
'Sellotape' or glue and brush
Autoradiography facilities (see Chapter 7)
Scissors
Measuring cylinders, 100 ml
Grease
Hot paraffin wax and paint brushes

THE DISTRIBUTION OF ^{32}P IN THE ANIMAL BODY

Although it is probably true to say that even a single radioactive molecule can be dangerous when it disintegrates, it is generally accepted that the danger from any particular isotope is proportional to its concentration. The maximum permissible quantity of any isotope used for experimental purposes, which can enter the environment and that can be ingested accidentally or intentionally, is controlled by Government regulations. However, as far as internal concentrations are concerned calculations based on an even distribution of the isotope throughout the body may be misleading since a number of isotopes are known to be concentrated in certain organs and tissues. Iodine is a notable example of an isotope showing uneven distribution and its accumulation in the thyroid gland has been used therapeutically in treating, for example, goitre. Calcium, of course, will concentrate in bone and in the following experiment the accumulation of ^{32}P in the bones of growing mice will be demonstrated.

During this experiment, it is seen that injected ^{32}P rapidly appears in the kidneys, as some of it is excreted in the urine, and then becomes generally distributed in the body tissues during the first few days. Subsequently, the ^{32}P level in most body tissues falls to zero while the concentration in bones rises to a very high level and then only gradually falls due to normal molecular turnover.

Procedure

Take six, young, healthy mice of the same sex, about 1-3 months old and inject about 5 μCi of [^{32}P]-phosphate into the muscle at the back of the neck. Use a sample of sufficiently high specific activity that the volume injected is fairly low (0·1 ml) and take the necessary precautions to avoid injecting air.

Place these mice in suitable cages, mark each with radioactive warning tape and place them so that all the litter can be collected separately and disposed of as radioactive waste. No radioactive gas will be emitted.

At suitable time intervals (for example 30 min, 1 day, 1 week, 2 weeks, 3 weeks and 1 month after inoculation) kill the animals and either process immediately or deep freeze in labelled polythene bags until later.

For each mouse accurately weigh six labelled small volume conical flasks and using plastic disposable gloves, dissect each animal in turn to provide:

(a) a piece of abdominal skin about 0·5 cm square
(b) about one-quarter of the animal's liver
(c) about one-quarter of its intestines
(d) a kidney
(e) the thigh muscle of a rear leg and
(f) the long bone of a rear leg—this should be scraped of muscle.

Remember that these animals are radioactive. Care should be taken to avoid contamination of yourself and the laboratory.

Place each tissue or organ into one of the conical flasks and reweigh to determine its weight. When all the flasks are ready add 3 ml concentrated nitric acid to each and heat gently for a few minutes on a warm hot plate to digest the tissue. Take care to avoid excess frothing and remove the flasks from the hot plate if this occurs. Carry out this procedure in a fume cupboard running at full extraction. As soon as digestion is complete remove the flasks from the hot plate and allow them to cool before decanting the contents into labelled counting tubes or vials.

Add 10 ml of 0·02% aqueous methyl umbelliferone solution to each vial and determine the activity of each vial, and of a background vial, using a scintillation counter and the Cerenkov technique. The efficiency of counting can be determined either by internal standardisation or by preparing a series of channels ratio correction curves using a suitable quantity of ^{32}P and a range of volumes of each tissue acid digest (see Chapter 3 and the references cited in the Bibliography for more details). Internal standardisation is perhaps the easiest method and is accomplished by accurately adding a small quantity of ^{32}P solution to each vial (including the background vial) and recounting them.

Measurement on a Geiger-Muller system

While ^{32}P present in the samples can be detected quite easily by Geiger-Muller counting, sample preparation can present a problem since the nitric acid attacks the aluminium planchets.

To overcome this problem the solution can be neutralised by 5 M Tris or another, usually more complex, digestion technique employed (see the references on scintillation counting). Alternatively, it is adequate in a student practical to evaporate 1 ml of each sample to dryness in a glass container (watch glass, small vial, etc.) and count this provided care is taken to obtain reproducible geometry. Duplicate or preferably triplicate counts will be necessary.

Calculation and expression of results
(a) *Scintillation counting*

This method assumes that all vials, including the background vial, have been internally standardised as described earlier.

Obtain a true count rate for each experimental sample (before the addition of ^{32}P) by subtracting the background count rate from each sample count rate.

In order to calculate the efficiency of counting ^{32}P in each experimental sample, assume that the additional count produced by the background vial after the addition of ^{32}P represents the true disintegration rate of the volume of ^{32}P added to the vials. Calculate the percentage efficiency of counting this added ^{32}P in each vial.

Thus, if counts in the background vial due to added ^{32}P =

X (i.e. total counts-counts before the addition)
and counts in the experimental vial due to the added ^{32}P=Y,
then the efficiency of counting the ^{32}P in the experiment =
$\dfrac{Y}{X} \times 100$

Having determined how efficiently the added ^{32}P is
counted in each experimental vial, use this efficiency value to
calculate the true disintegration rate of the ^{32}P originally
present in the experimental material in each vial.

Thus, if the experimental sample gave an original count of
Z then the true disintegration rate is

$$Z \times \frac{100}{\% \text{ efficiency}}$$

Plot graphs of the activities mg^{-1} wet weight against time
after injection for each of the tissues samples for all mice.

(b) Geiger-Muller counting

Average the replicate counts for each sample and subtract
the average background count rate from each experimental
sample.

Plot graphs as described above.

If the mice have been processed on different days the
results will have to be corrected for the decay of ^{32}P
occurring during the time elapsed from the day of processing
the first mouse. Since it is the distribution of ^{32}P within the
animal body that is of interest it is not necessary to calculate
the actual quantities of ^{32}P present in each sample and a
decay correction back to the date and time of inoculation is
unnecessary.

Waste disposal

The liquid samples from this experiment can be disposed
to the drain as usual. It is necessary to collect all the animal
carcasses and waste tissue together with the litter from the
cages into a polythene bag for disposal according to standard
procedures.

Student level
HNC to Degree students.

Time required

The full experiment lasts about 1 month with the mice being assayed at periodic intervals. However, it is more convenient for a technician to kill and deep freeze the mice at various times and to present the students with these mice. The experiment can then be completed within three hours if each group of students is not asked to process all six mice.

Materials required

	Per group		
Mice from the same litter, 1-3 months old and of the same sex	6	flasks	6 per mouse
Animal cage	1	Marking pen	1
30 μCi [^{32}P]-phosphate in a fine needle calibrated hypodermic syringe for injection	1	Concentrated nitric acid	20 ml per mouse
		5 ml safety pipette	1
		Adjustable hot plate	1
Dissection instruments; 2 awls, scissors, scalpel and fine forceps		Tongs	1 pair
		Test tubes in rack	7 per mouse
Dissection board with several thicknesses of paper towel pinned to it to reduce contamination of the wood	1	Scintillation vials or planchets	7 per mouse
		For general use	
Disposable plastic gloves	1 pair	Accurate balances	
Polythene bags	6	Fume cupboards	
25 ml or 50 ml conical		Deep freeze	

THE ABSORPTION OF [^{14}C]-L-GLUTAMATE BY RAT INTESTINE

The main function of the intestine of animals is to digest dietary food materials. However, before the products of digestion are available to the animal they must be absorbed from the contents of the gut and the intestinal cells are modified to absorb a wide range of dietary products by active transport processes.

This transport is unilateral, occurring from inside of the gut to the surrounding blood capillaries, but by everting the intestine and forming a sac, materials will be transported into the sac and can be collected and analysed.

In the following experiment, the transport of $[^{14}C]$-L-glutamate by the intestine can be measured and to determine whether it is occurring by active transport, the effect of cyanide can be examined.

Procedure

1. *Preparation of intestine sacs*

It is advisable to note that while intestines are fairly robust and should retain their physiological activities for some time it is as well to take certain precautions to extend their life. In particular, they should never be allowed to dry out, they should be handled carefully and as far as possible only by the cut ends, and at all costs cutting of the intestine wall should be avoided. Some workers prefer to use hamster rather than rat intestine because of its longer life and greater physical strength.

Kill a rat by a sharp blow on the head and pin it, on its back, to a dissection board. Open the rat by an incision up the abdomen as far as the thorax and cut laterally to open the cavity completely. Remove and discard the liver. Cut the duodenum off close to the stomach, insert a blunt syringe needle or cannula tube and tie it tightly with thread. Using a hypodermic syringe slowly pump warm oxygenated saline-glucose through the intestine until it appears in the caecum (15–25 ml). This should wash out most of the food material. Make a second cut where the ileum joins the caecum and force more saline-glucose through the gut to wash it out completely. Holding the lowest portion of the ileum carefully remove the gut from the mesenteries. Take care to keep the gut wet with saline-glucose and not to damage it. Cut the intestine at the base of the duodenum and remove it to a Petri dish of saline-glucose.

Tie a thread round one of the cut ends of the intestine and using a smooth taper-tipped glass rod of smaller diameter than the intestine, push this tied end into the cavity of the intestine. Using hands which have been wetted with saline-glucose gently ease the rest of the intestine back down the rod until the invaginated end appears out of the open end of the intestine. Complete the inversion of the intestine, remove it from the rod, cut it in half and immerse the halves in a dish of saline-glucose.

2. Setting up the experiment

Around one end of one of the lengths of gut tie two separate pieces of thread and into the other end insert a blunt syringe needle or cannula tube. Tie a thread round the gut to hold the needle or cannula in place and inject a measured volume (at least 1 ml) of Krebs-Ringer into the sac together with a large air bubble. Remove the needle or cannula, pull the thread tight and tie another thread at this end. Leakage into the gut sac from the ends must be avoided.

Place the sac into a small conical flask containing 50 ml of 10 mM glucose and 5 ml glutamate containing 5 μC_i [^{14}C]-L-glutamate. Clamp the flask into a water bath at 37°C and gently aerate for one hour. Repeat this procedure using the other piece of intestine, injecting the same volume of Krebs-Ringer but place this sac into a flask containing 10 mM glucose, 10—^{3} M NaCN and 5 ml glutamate containing 5μC_i[^{14}C]-glutamate.

3. Preparation and counting of experimental samples

At the end of the experimental period, pour off most of the solution into a beaker and decant the gut sac into a Petri dish. Pick up the sac by one end using a pair of forceps and wipe the sides gently with the edge of a filter paper before rinsing it by dipping into a beaker of Ringer. Hold the sac just inside a centrifuge tube, cut the end and allow the contents to drain. Centrifuge at full speed on a bench centrifuge for two minutes to clarify the liquid and remove the supernatant with a Pasteur pipette.

Using Pasteur pipettes transfer equal numbers of drops of each sample on to two labelled planchets and on to two other planchets place the same number of drops of the reaction mixture from the flask. Prepare four similar planchets from the cyanide containing experimental system. Add two drops of dilute detergent, evaporate and count each planchet twice for 100 sec. Alternatively, 0·2 ml samples can be counted by scintillation counting.

Tabulate the actual counts obtained and the means of these counts before and after background subtraction.

Comment on and explain all the results.

Student level
HNC and post first-year undergraduates.

Time required
2–3 h.

Materials required

Per group

Rat—this should be starved for 18-24 h and freshly killed by non-chemical means	1	
Dissecting board, awls and dissection guide		
Dissecting instruments including scissors and fine and blunt forceps		
Hypodermic syringes 5 and 25 ml	1	
Syringe needles blunted on a carborundum stone	2	
Petri dishes	3	
Glass rod 30 cm long pulled out to a smooth cone-shaped tip	1	
Beakers	2	
10 ml centrifuge tubes	2	
Aquarium pumped linked via rubber tubing and a 'Y' piece to 2 Pasteur pipettes and 2 screw clamps	1	
Clamp stand with 2 bosses	1	
Saline-glucose (0·9% NaCl + 0·3% glucose) kept at 37°C and aerated for 1 h before the experiment	200 ml	

Krebs-Henseleit Ringer treated as above   10 ml

150 ml 0·9% NaCl
50 ml 0·154 M KCl
315 mL 0·15 M $NaHCO_3$ } dilute to 1,500 ml with water
15 ml 0·15 M KH_2PO_4
15 ml 0·15 M $MgSO_4$

Glucose (10 mM) + Glutamate (5 mM) containing 5 μCi of [^{14}C]-L-glutamate
Glucose-glutamate as above with NaCN (10^{-3} M)

50 ml

For general use
Fine cannula tubing
Strong thread
Filter paper
Water baths at 37°C
Pasteur pipettes and teats
Test tubes
Planchets, dilute detergent and heat lamps or scintillation vials and scintillant
G-M or scintillation counting system.

STUDIES ON SODIUM TRANSPORT ACROSS FROG
SKIN USING [22]Na

The structure and function of cell membranes has been of interest for a considerable number of years and a wide range of experimental methods and biological materials have been used in their investigation. The movement of sodium and potassium ions across animal cell membranes has been of particular interest because of its fundamental importance in a number of situations, for example, nerve impulse conduction and kidney tubule and red blood cell function. The necessity to keep isolated animal components in solutions containing sodium in order to maintain maximum physiological function has led to the widespread use of radioactive forms of sodium as tracers for following the movement of added sodium from one compartment to another when both already contain sodium.

Both of the commercially available radioisotopes of sodium are reasonably powerful γ-emitters and hence need to be treated with much more respect than the other radioisotopes included in experiments described in this book. Two radioactive sodium experiments have nevertheless been included because they illustrate important aspects of modern biology, yet are comparatively easy to perform. In both cases, [22]Na ($t_{\frac{1}{2}}$ = 2·6 yrs) is used in preference to [24]Na ($t_{\frac{1}{2}}$ = 15 h) because of its much longer half life and lower energy β and γ-emissions.

The common frog produces copious quantities of urine and as a consequence loses considerable amounts of sodium. This loss is compensated for by the uptake of sodium ions from the surrounding water and in the experiment described below, sodium movement across isolated abdominal skin will be demonstrated. Variation of temperature, starvation of the isolated skin followed by glucose addition or the effect of respiratory inhibitors can be used to demonstrate the active nature of this process.

Procedure

Kill a frog by some non-chemical means and wash the animal thoroughly in cold tap water. Lay it on its back and remove the loose abdominal skin in a single piece. Tie the skin securely over the smooth end of a 2 cm diameter glass test tube (cut to about 10 cm length and flame polished) with the internal side of the skin facing the inside of the tube.

Clamp the tube within a 10 ml beaker containing 3 ml of frog Ringer's solution and add 2 ml of Ringer's solution to the inside of the tube. By adjusting the position of the tube equalise the fluid levels on both sides of the tube in order to prevent hydrostatic pressure forcing solution through the membrane. Bubble air through the two compartments at a rate of 2 or 3 bubbles per second.

If possible build a lead wall in front of, and preferably all round the apparatus for protection against the γ-rays. In any case, stand well away from the apparatus during the intervals between sampling.

Allow the system to stabilise for 10 min and then add 5 μCi of ^{22}Na solution to the outer compartment and an identical volume of distilled water to the inner compartment. Withdraw 0·05 ml of the inner solution to a scintillation vial containing 5 ml of scintillation fluid or dry it on to a planchet using a heat lamp. Replace the volume taken with 0·05 ml of Ringer's solution.

Repeat this procedure every 10 min for 50 min.

After the sixth sample has been taken replace the volume removed with 0·05 ml of 0·5 M KCN solution and continue taking samples for a further 50 min replacing the volume taken on each occasion with 0·05 ml of Ringer's solution.

Counting and expression of results

Obtain duplicate count rate determinations for each planchet or scintillation vial and calculate their means. Subtract a mean background count rate and plot the corrected experimental count rates against time.

Student level
2nd- and 3rd-year degree students.

Time required
The whole experiment can be completed within 3 h.

Materials required

	Per group
Air pump with two narrow outlets, e.g. tips from Pasteur pipettes or hypodermic syringe needles	
Clock	
Clamp stand	
Cotton, scissors	
Micro-pipette, 50 μl	1
2 cm diameter test tube, cut to about 10 cm length and flame polished	1
10 ml beaker	1
Frog (kept in distilled water for two days)	1
Plastic scintillation vials (and 60 ml scintillant) or planchets (and a heat lamp)	12
KCN 0·5 M	50 μl
^{22}Na solution	5 μCi
Frog Ringer's solution (pH 7·0-7·5):	10 ml
NaCl	0·65 gl^{-1}
KCl	0·025 gl^{-1}
CaCl$_2$ (anhydrous)	0·03 gl^{-1}
NaHCO$_3$	0·02 gl^{-1}

STUDIES ON SODIUM EXCHANGE IN CRUSTACEANS USING ^{22}Na

For a brief discussion of the use of radioisotopes in studies on sodium ion movement see the previous experiment.

Any aquatic animal is continually exchanging ions present in its body with similar ions present in the environment. A dynamic equilibrium develops in which there may be a difference in the concentration of an ion between the inside of the animal and the environment and there is a continual movement of ions in both directions.

In these experiments, the movement of sodium in both directions can be measured using the radioisotope ^{22}Na.

Interesting results are obtained if animals are placed in alien environments (e.g. edible crabs in fresh water,.etc.) and the rates of sodium movement compared with the rates obtained when they are in their normal environments.

EXPERIMENT 1. SODIUM INFLUX

Carefully tie a piece of cotton around the carapace of a crab or crayfish and place it in a beaker of sea-water or fresh water (containing $0 \cdot 05$ μCi of ^{22}Na ml^{-1}). Leave it in this solution for 1 h and during this time weigh five scintillation vials or planchets.

Use the cotton to remove the crustacean and pass it successively through four beakers of water for about 15 sec each in order to remove surface contamination. Use a hypodermic syringe to remove a small blood sample (about $0 \cdot 1$ ml) from the membrane at the joint of a leg or at the junction of the carapace and abdomen. Replace the crab in the active solution.

Eject the blood into a scintillation vial or planchet, reweigh it and find the weight of the sample by subtraction. Take, and process, further samples at hourly intervals and preferably one sample late in the afternoon for a morning exercise or the following morning for an afternoon exercise.

Add 5 ml of scintillant or dry the samples on to their planchets with a heat lamp and determine the radioactivity on a scintillation counter or Geiger-Muller system. Subtract the background count rate obtained from a vial of scintillant or an empty planchet and plot counts min^{-1} mg^{-1} sample against time of withdrawal of the sample.

EXPERIMENT 2. SODIUM EFFLUX

Carefully tie a piece of cotton around the carapace of a crustacean and place it in a solution appropriate to its natural environment but which contains $0 \cdot 05$ μCi of ^{22}Na ml^{-1}. Leave overnight for the animal to reach an equilibrium with the environment.

Remove the crustacean and wash it by dipping for 15 sec on each occasion in four changes of water. Place it in a large beaker and add sufficient sea or fresh water to cover the animal. Immediately remove 1 ml of this water to a scintillation vial or planchet and replace it with 1 ml of sea-water or fresh water.

Take further samples at 30 min intervals replacing the volume taken on each occasion by 1 ml of appropriate solution.

Add 5 ml of a suitable scintillant or dry them on to their planchets with a heat lamp. Determine the activities in a scintillation counter or on a Geiger-Muller system and subtract the background count rate obtained from 1 ml of bathing medium. Plot a graph of count rate against time.

Student level
2nd- and 3rd-year degree students.

Time required
Sodium influx: 3 h (minimum)
Sodium efflux: equilibration (overnight)
　　　　　　　experiment (3 h—can be run concurrently with the sodium influx experiment)

Materials required

	Per group		
Sodium influx		*Sodium efflux*	
500 ml beakers	5	500 ml beakers	5
1 ml hypodermic syringe	1	1 ml hypodermic syringe	1
Sea-water or fresh water containing $0 \cdot 05$ μCi ^{22}Na ml^{-1}	100 ml	Sea-water or fresh water	410 ml
		Sea-water or fresh water containing $0 \cdot 05$ μCi ^{22}Na ml^{-1}	100 ml
Scintillation vials (and 30 ml scintillant) or planchets (and a heat lamp)	6	Scintillation vials (and 30 ml scintillant) or planchets (and a heat lamp)	6
Cotton		Cotton	
Clock		Clock	
For general use		*For general use*	
Balances		Crustacea	
Crustacea			

DETERMINATION OF DROSOPHILA POPULATION NUMBERS

If a small number of radioactive flies are released into a much larger number of non-radioactive ones, the radioactivity becomes diluted by a factor which depends upon the total number of flies present. If the radioactivity is subsequently determined in random samples of the flies it is possible to determine the number of flies in the original population using the isotope dilution principle.

Drosophila can be cultured on a supplemented agar medium with the larvae feeding on yeast cells inoculated into this medium. If the yeast cells are made radioactive by allowing them to grow before inoculation on a medium containing ^{32}P, then the larvae and subsequently the flies become radioactive. These flies can then be etherised and released into the larger population of non-radioactive flies.

Procedure

The complete experiment is described below. Students can either carry out the entire experiment over several weeks or be presented with the *Drosophila* cultures and be asked to carry out the final estimation.

1. *Preparation of the medium*

Measure 40 ml of black treacle into a flask and add 100 ml of hot water, including the rinsings from the measuring cylinder. Add 5 g of agar and shake well before autoclaving at 15 lb/sq in pressure for 30 min.

Weigh 40 g maize meal into a litre beaker and add 340 ml of water. Heat the beaker in a water bath for 30 min at 100°C, keeping it well stirred.

To the black treacle/agar mixture add 100 mg methyl-p-hydroxybenzoate or a commercial anti-fungal agent such as 'Nipagin', and then pour this mixture slowly and with constant stirring into the maize meal mash.

Distribute the medium between five sterilised 1 pt milk bottles or five 500 ml conical flasks and plug each with sterile non-absorbent cotton wool. Allow the medium to cool.

Certain biological supply houses market prepared dried medium and this may prove convenient.

2. *Preparation of the yeast*

Into a 100 ml conical flask measure the following:

3 ml 0·5 M glucose

6 ml 0·05 M phosphate buffer pH 7 containing 15 μCi ^{32}P

9 ml water

2 ml 1% yeast suspension.

Incubate this culture at 37°C overnight, preferably in a shaking water bath. Prepare a similar culture without ^{32}P for each bottle of non-radioactive flies to be grown.

Centrifuge the yeast cells at low speed, remove most of the supernatant to a waste bottle and then resuspend the cells in minimum volumes of supernatant using Pasteur pipettes. Add all of the radioactive cells to one of the milk bottles or conical flasks making sure that the suspension is placed on the agar. Allow the bottle at least an hour at 37°C to set and then add a small piece of sterile filter paper to the vessel by pushing the bottom of it vertically into the medium using a sterile wire. Treat each of the other containers with non-radioactive yeast suspension and filter paper in a similar manner.

3. *Inoculation*

To all containers add about ten to twelve wild-type *Drosophila* flies and incubate them at 25°C until a large number of eggs are visible. Remove the flies and incubate the containers until a large number of flies are present (about two weeks).

4. *Population estimation*

In order that the determination of radioactivity should be as easy as possible and to stop the possibility of radioactive flies escaping into the laboratory, all flies should be killed before mixing and counting.

Pour ether into the cotton wool plug of each container, invert it and leave it until all the flies are dead. Remove as many dead flies as possible from each container into a Petri dish. Separately weigh ten of the non-radioactive flies and then all of the non-radioactive flies.

Count the total number of radioactive flies by gently pushing them to one side of the dish with a paint brush. Remove about five to ten flies on to each of five planchets and cover each planchet with 'Sellotape' making sure the flies do not stick to the adhesive. Count each planchet twice using a long time interval (300 sec) and after averaging the count rates, determine the count rate per fly.

Remove the flies from the planchets and tip all the radio-active flies into a clean and dry litre flask. Add the non-radioactive flies and shake well to mix them. Tip the mixture on to a large Petri dish and count twenty-five to thirty flies on to each of four or five planchets. Cover each with 'Sellotape' as before and count each planchet twice using a long time interval.

Average the count rate for all the planchets and determine the average number of radioactive flies present. From this determine the dilution of the isotope and hence the total number of flies present.

From the weight of the ten non-radioactive flies and the weight of the total population one can determine the number of flies in the non-radioactive population and use this as a check on the results obtained by the isotope dilution method.

Student level
HNC to degree students

Time required
Preparation and inoculation of cultures 3–4 h
Growth of *Drosophila* cultures (at 25°C) 2 wks
Population determination 2–3 h.

Materials required

	Per group		
Black treacle	40 ml	Measuring cylinders 500	
Agar	5 g	ml, 100 ml, 50 ml	1 of
Maize meal	40 g		each
Methyl-*p*-hydroxybenzo-		1 litre beaker	1
ate or 'Nipagin'	100 mg	Stirring rod	1
0·5 M glucose	3 ml	1 pt milk bottles or 500 ml	
0·05 M phosphate buffer		conical flasks	5
pH 7 + 15 μC$_i$^{32}P	6 ml	Conical flask 100 ml	1
1% yeast	1 ml	Small paint brush	1

For general use
Autoclave or pressure cooker
Water bath at 100°C
Ethanol
Ether
Non-absorbent cotton wool
Centrifuge and tubes
Radioactive waste bottle

Pasteur pipettes and teats
Sterile filter paper discs (2-3 cm diameter)
Culture of wild-type *Drosophila*
Incubator at 20-30°C
Petri dishes
Planchets
End window counter

ENVIRONMENTAL RADIATION

It is quite easy to demonstrate the existence of various sources of environmental ionising radiation. This radiation contributes quite significantly to the normal mutation rate of humans and measurement of it provides a useful way of introducing students to the Geiger-Muller counter and allowing very young students to do some work with isotopes and radiation.

Background radiation

This experiment is intended to give an indication of the amount of radiation to which organisms are continually exposed. Most of the radiation measured in this experiment is cosmic radiation and it is well to remember that there will be other forms of natural radiation in addition.

If a suitable Geiger-Muller system is not set up for you, do so yourself according to the instructions given and allow it to stabilise for 10 min. Measure the count rate for three periods of 100 or 60 sec and determine the mean. Given that the Geiger-Muller tube has a diameter of 2·5 cm and a length of 10 cm calculate its surface area from the formula:

$$2\pi rh + 2\pi r^2 \text{ (r is the radius, } 1\cdot 25 \text{ cm, and h the length, } 10 \text{ cm)}$$

Calculate the counts produced per second per cm² of tube surface area; calculate your body surface area using the nomogram supplied (Figure 25) and determine the background irradiation over the whole body surface over a 24 h period.

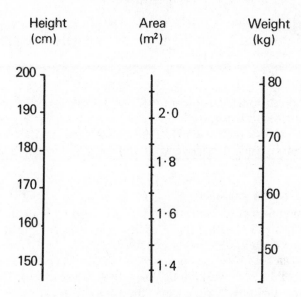

Use a ruler to link naked body weight and height—the value on the centre scale is the body area in square metres.
Note that 1 pound weight is equal to 0·45 kg and that 1 inch is equal to 2·54 cm.

Figure 25. Nomogram for calculating body surface area.

Bear in mind that since the efficiency of counting is not 100% and the radiation counted is only that which passed through the lead shielding, the actual incidence of natural radiation must be very much higher than your calculated figure.

Air filtration
It is quite surprising to find that the atmosphere contains significant quantities of radioactive particles. This is not the cosmic radiation from space measured in the previous experiment but is due to isotopic materials absorbed into and on, particles of atmospheric dust. Some of these isotopes arise due to interaction of cosmic radiation and atmospheric

materials but others are fission products from nuclear explosions.. The presence of these materials can produce measurable counts in rain.

You will be supplied with filter paper through which air has been sucked for a period of hours. Cut a piece of this about 2 cm square and count it for 100 or 60 sec on shelf 1 of a Geiger-Muller system. If you are told the air flow rate and area of the paper, calculate the concentration of isotopic material in the atmosphere assuming 100% efficiency of counting.

Laboratory chemicals

Many naturally occurring chemicals contain a proportion of radioactive nuclides of their elements—thus some general laboratory chemicals are radioactive.

Spread a thin layer of each of the salts provided onto planchets, cover with 'Sellotape' and determine the count rate produced in each case. Compare these rates with the background count rate obtained by counting an empty planchet.

Student level
All secondary school students and above.

Time required
45-60 min.

Materials required

Air filtration. An air filtration system such as a Buchner flask and funnel covered with filter paper sealed at the edges with 'Sellotape'. Air is sucked through this for 12-24 h, preferably at a known flow rate.

Scissors
Laboratory chemicals
'Sellotape'
Spatula
Salts of uranium, thorium, lead and potassium

All experiments
Geiger-Muller   counters   and planchets

THE CHARACTERISTICS OF RADIOISOTOPE COUNTING APPARATUS AND THE RADIATIONS EMITTED FROM ^{3}H, ^{14}C AND ^{32}P

In many courses of practical radiobiology it is felt desirable to include a few simple experiments to illustrate the characteristics of the apparatus used to count radioisotopes and even in lower level or otherwise different courses such experiments are useful in providing a means by which students can be introduced to radioisotopes.

A few such experiments for both the Geiger-Muller and scintillation counters are described below.

(1) DEMONSTRATION OF THE GEIGER-MULLER COUNTER PLATEAU

The characteristic features of Geiger-Muller tubes and their operation were discussed in the introductory chapters. It is essential that the tube be operated at an acceptable voltage and this has to be determined for each tube.

The experiment described here concerns the method of determining this voltage and also illustrates the principle and method of operation of analytical Geiger counters.

The experiment is used to determine the starting and threshold voltages and the shape of the plateau. From these data the optimum operating voltage and plateau slope can be calculated.

Source preparation

In experiments of this type it is useful to have a reference source of constant activity by which counting over a period of time may be standardised.

Uranium oxide (U_3O_8) which has not been chemically treated for at least one year is particularly suitable as it combines chemical stability with long half-life ($4 \cdot 5 \times 10^9$ years). A slurry is made from uranium oxide, acetone and a small amount of adhesive in a disposable plastic beaker. Small amounts are transferred by Pasteur pipette or a glass rod to a planchet, spread evenly and dried under a heat lamp ensuring that the acetone does not boil. It is best to filter out all the emissions except the β-particle of 2·31 MeV energy by covering the planchet with aluminium foil of thickness equivalent to 54 mg cm^{-2} (0·0008"). Domestic cooking foil weighs about 5 mg cm^{-2}.

The planchet should be counted for a short time and if a count rate of 15,000-20,000 counts 100 sec^{-1} is not achieved the foil should be removed and additional layers of U_3O_8 should be built up by adding small amounts of the slurry, with drying and counting between the additions. Finally, seal the aluminium foil by smearing adhesive round the planchet edge and label the planchet with its contents and owner's initials.

As an alternative to the above the following experiment could be performed using other sources, preferably of isotopes emitting relatively powerful β-particles, e.g. ^{32}P. This, of course, has the disadvantage of rather rapid decay although this is not short enough to seriously affect the validity of the experiment.

Counting equipment

The equipment required for the experiment is the standard Geiger-Muller tube assembly enclosed in a shelf-type lead castle connected through a probe unit to a power unit and scaler. Most commonly these are integrated into a single assembly. The equipment can either be set up for the student or he can be allowed to do so himself under careful supervision.

Procedure

After setting up the equipment and allowing it to stabilise, the source to be used should be placed in the lead castle using shelf number two for U_3O_8 and shelf number one for other, weaker sources.

The operating voltage should be set at 0 V. To determine the starting potential (V_s) the voltage should be slowly raised until the scaler begins to register pulses. Turn the voltage down 20 V and *very slowly* raise it until counts are registered. This will give a fairly accurate indication of the starting voltage.

To determine the plateau slope the source should be counted at V_s for 100 sec. The voltage should then be increased in stages of 25 V up to V_s + 100 V and counts over a 100 sec period noted in each case. If at any time the counter begins to race the voltage should be immediately turned down to prevent tube destruction.

Results and calculations

Tabulate EHT voltages against counts per 100 sec after correction for the paralysis time of the counter (Appendix 2) and plot corrected counts against voltage.

The plateau threshold voltage (V_T) is the voltage at which the linear portion of the graph (the plateau) begins. The plateau slope can be calculated from the equation:

$$\frac{C_2 - C_1}{C_M} \times \frac{100}{V_2 - V_1}$$

where C_2 and C_1 are two count values on the linear portion of the plateau (C_2 being greater than C_1), C_M is their mean and V_2 and V_1 their voltages.

The slope is usually less than 0.05% per volt and may be negligible. As the Geiger-Muller tube ages the slope increases and shortens. To allow for this, the normal operating voltage is chosen as V_T + 100 V, i.e. in the lower one-third of the plateau.

(2) COMPARISON OF SCINTILLATION COUNTING AND GEIGER COUNTING

Take two planchets and mark the underside ^3H and ^{14}C, respectively. Place 0.1 ml of the appropriate isotope solution on its planchet and add one drop of dilute detergent. Dry the two planchets under a heat lamp. Determine the radioactivity by counting each planchet for 300 sec on a Geiger-Muller system. Determine the background count rate by counting a clean planchet.

Take two plastic scintillation vials and mark the caps ^3H and ^{14}C. Place 0.1 ml of the appropriate isotope solution in its vial and add 3 ml of the scintillant provided. Place in the scintillation counter, record the number of their belt positions and count them according to the instructions provided. A background vial is provided for you to count.

Compare the counts per minute obtained for each sample and comment on the results.

(3) A COMPARISON BETWEEN SCINTILLATION COUNTING AND CERENKOV COUNTING

Take six scintillation bottles; in two, place 100 ml dioxan-based scintillant, in another two, place 100 ml water and in the remaining two, place 100 ml 0·02% aqueous methyl umbelliferone solution. Carefully add 1 ml of the ^{32}P solution (about 0·01 μCi) provided, add numbered caps, mix well and place in the scintillation counter.

Determine the radioactivity in the bottles and prepare a table of your results.

(4) DETERMINATION OF EFFICIENCY OF SCINTILLATION COUNTING OF QUENCHED SAMPLES

Take two scintillation vials and mark the cap of one 'Q' and the cap of the other '−Q'. To each, add 3 ml of the scintillant provided. To the former, add 0·06 ml of carbon tetrachloride and to the latter 0·06 ml of water. To both add very carefully 0·1 ml of the ^{3}H solution used earlier and mix well and gently. Place in the scintillation counter, cool if necessary, and determine their activity.

Now when you have done this, you should find that the vial 'Q' gives a lower count rate than the other despite the fact that they will contain the same amount of radioactive material. This is because the CCl_4 acts as a 'quenching' agent and reduces the efficiency of energy transfer through the scintillant plus sample mixture.

The problem will therefore arise in any experimental sample; what is the efficiency of counting of the isotope and hence what is the true count rate? Several methods are available for determining the counting efficiency and two of them are described here:

(a) The method described in this section is known as the

sample channels ratio method. It is only suitable for scintillation counters with two counting channels.

Take seven scintillation vials (labelled 1-7) and add the following reagents to them (ml):

	1	2	3	4	5	6	7
Scintillant	3	3	3	3	3	3	3
Water	0·1	0·09	0·07	0·05	0·03	0·01	0
CCl_4	0	0·01	0·03	0·05	0·07	0·09	0·1

Very carefully add 0·1 ml of the ^3H isotope to each bottle and place them all in the scintillation counter. Count each vial for 1 min or 100 sec.

From the results:

(i) Prepare a calibration curve for the seven samples by plotting the channels ratio (ratio of counts per minute in channel A to counts per minute in B) against the efficiency (counts per minute in channel B divided by the disintegration rate for the isotope sample used). You will be told the disintegration rate.

(ii) Determine the channels ratio for the two samples ('Q' and '−Q') prepared before and using the calibration curve you have just made, determine the efficiency of counting and hence the true count rate.

(iii) Determine also the percentage of the total count which is present in each of channels A and B. Plot this percentage against volume of CCl_4 added.

(b) The method described in this section is termed internal standardisation. It is potentially the most accurate way of determining efficiency but is usually subject to some errors. It involves counting each sample twice and alters the original activity of the samples, thus preventing one from recounting them subsequently.

Very carefully add 0·1 ml of the ^3H isotope to both the 'Q' and '−Q' vials and also to a vial containing 3 ml of scintillant.

Assuming that the ^3H is counted with 100% efficiency in the vial containing scintillant only, calculate the percentage efficiency of counting this added ^3H in the vials 'Q' and '−Q'. Use these efficiency values to correct the count rate obtained for the original sample to its true disintegration rate.

(5) THE NATURE OF ^{3}H, ^{14}C AND ^{32}P EMISSIONS

It is possible to carry out in a comparatively short time several experiments which illustrate very well some of the important characteristics of the radiations emitted by the three biologically most important radioisotopes. These experiments are useful in that they illustrate some of the problems involved in working with these radioisotopes.

(a) The strength of the emitted radiation

While ^{32}P has a relatively strong β-emission (1·710 MeV max) that of ^{14}C is quite weak (0·156 MeV max) and ^{3}H weaker still (0·019 MeV max). This means that although ^{32}P can be counted fairly readily with reasonable efficiency on Geiger-Muller tubes the counting of ^{14}C and ^{3}H is seriously effected by such factors as self-absorption in thick samples, distance from the Geiger-Muller tube window, etc.

(i) *The effect of distance from the Geiger-Muller tube window.* Take three labelled planchets and pipette a small volume of solutions containing the same activity (about 1 μCi) of non-volatile compounds containing ^{32}P, ^{14}C and ^{3}H. Add 0·4 ml 1% detergent to produce an even film and dry under a heat lamp.

Count each planchet twice for 100 sec on each shelf of the lead castle. Measure the background count over 100 sec. Calculate the average count at each shelf height, subtract the background count and calculate the efficiency of counting for each shelf assuming shelf 1 is 100% efficient for ^{32}P. Plot a graph of efficiency or count rate against shelf height.

A very good indication of the comparative strengths of the emissions of these isotopes can be obtained if a laboratory contamination monitor is available capable of detecting ^{14}C.

Adjust the monitor to its maximum range and switch to the audio output. Place the end window of the detector about 2 cm above a planchet containing a dried sample of one of the isotopes (the planchets prepared in previous experiments are suitable). By slowly lifting the detector vertically above the planchet the rapid fall-off in count rate obtained with the ^{14}C sample, the very slow fall-off for ^{32}P and the inability to detect ^{3}H at all are quite marked.

(ii) Barrier attenuation. The weaker emission of ^{14}C is readily filtered out by barriers, even those as thin as a piece of 'Sellotape'. This can have serious consequences if particulate material is to be counted and the planchet is covered for the sake of safety.

Using the ^{14}C and ^{32}P planchets prepared for the previous experiment, cover each with a single piece of wide 'Sellotape' and count again on shelf 1 of the lead castle. Repeat with two and three 'Sellotape' strips over the planchet.

Calculate the percentage attenuation in each case.

(iii) Self-absorption in thick samples. In biological experiments it is often necessary to measure the activity of samples containing appreciable amounts of solid material. These materials usually show a reduced count rate due to the emitted radiation being absorbed by the sample (self-absorption) and while it is desirable in comparative experiments to prepare planchets containing equal weights of sample the efficiency of counting can still be greatly reduced. This is particularly the case for weak β-emitters such as ^{14}C but is still appreciable for ^{32}P.

Make up solutions of ^{32}P and $0 \cdot 1\%$ NaCl and a ^{14}C compound and $0 \cdot 1\%$ NaCl to give about $1 \mu Ci$ ml$-^1$. Pipette $0 \cdot 1$ ml of each on to separate planchets, add two drops of detergent, dry the planchets and count them twice for 100 sec using shelf 1.

Add a further $0 \cdot 1$ ml of each solution and detergent, dry and recount. Repeat this procedure until a total of 1 ml has been added.

Tabulate the actual and average count rates in each case, the expected count rate based on the first count obtained and calculate the percentage efficiency of counting for each volume addition of sample.

(b) The half-life of ^{14}C and ^{32}P

A particular problem associated with the use of ^{32}P and a number of other isotopes is their short half-life. It is not possible to store these isotopes for long periods or to use them in long-term experiments unless suitable corrections can be made. In very accurate work, corrections due to the decay of ^{32}P during the course of counting a large number of samples have to be made.

Prepare two sources by adding 0·5 ml of ^{32}P and ^{14}C-labelled substrate solutions to separate labelled planchets, adding one drop of dilute detergent solution and drying under a heat lamp. The solutions should contain about 2 μCi ml^{-1}.

Count each sample three times for 100 sec on shelf 1 of a lead castle. Repeat the sample counting as often as possible over a period of a week or two, using the same counting system and counting geometry if at all possible. Note the precise time of counting in each case.

Plot counts per 100 sec (using corrected counts if possible) against time from sample preparation in days and hours. Use semi-log graph paper in order to obtain a straight line.

The half-life of the isotopes can be calculated in two ways.
(i) from the graph itself (even if this involves extrapolation) the time taken for the count-rate to fall to 50% of a previous value can be measured.
(ii) If the plot has been made on a semi-logarithmic scale then the decay constant (λ) can be calculated from the slope of the graph.

$$\text{Slope} = - \frac{\lambda}{2 \cdot 303}$$

The half-life $(t_{\frac{1}{2}})$ can be calculated from the decay constant using the equation:

$$t_{\frac{1}{2}} = \frac{0 \cdot 693}{\lambda}$$

The half-life of ^{14}C is so long that it will not be possible to calculate it in this experiment using either method.

Appendices

APPENDIX 1

The Properties of some Ionising Radiations

Type	Mass Ratio (Relative)	Energy (Relative)	Penetration (Relative)	Ionisation (Relative)
Alpha	7380	Moderate	1	10^4
Beta	1	Low-high	10^2	10^2
Gamma	0	High	10^3-10^4	1

APPENDIX 2

Table of Lost Counts for a Geiger-Muller System with a Paralysis Time of 400 Microseconds

Lost counts are calculated from the formula:

$$\left(N \times \frac{t}{t - (N \times 4 \times 10^{-4})} \right) - N$$

where N is the number of counts obtained in time t using a Geiger-Muller system of dead time 4×10^{-4} sec.

Column A.   Experimental count rate obtained.
Column B.   Counts to be added if column A is counts min^{-1}.
Column C.   Counts to be added if column A is counts 100 sec^{-1}.

A	B	C	A	B	C
50	0	0	15050	1678	964
100	0	0	15100	1690	970
150	0	0	15150	1702	977
200	0	0	15200	1713	983
250	0	0	15250	1725	990
300	0	0	15300	1737	997
350	0	0	15350	1749	1004
400	1	0	15400	1761	1010
450	1	0	15450	1774	1017
500	1	1	15500	1786	1024
550	2	1	15550	1798	1031
600	2	1	15600	1810	1038

A	B	C	A	B	C
650	2	1	15650	1823	1045
700	3	1	15700	1835	1052
750	3	2	15750	1847	1058
800	4	2	15800	1860	1065
850	4	2	15850	1872	1072
900	5	3	15900	1885	1079
950	6	3	15950	1897	1086
1000	6	4	16000	1910	1094
1050	7	4	16050	1923	1101
1100	8	4	16100	1935	1108
1150	8	5	16150	1948	1115
1200	9	5	16200	1961	1121
1250	10	6	16250	1974	1129
1300	11	6	16300	1987	1136
1350	12	7	16350	2000	1144
1400	13	7	16400	2013	1151
1450	14	8	16450	2026	1158
1500	15	9	16500	2039	1165
1550	16	9	16550	2052	1173
1600	17	10	16600	2065	1180
1650	18	10	16650	2078	1188
1700	19	11	16700	2092	1195
1750	20	12	16750	2105	1202
1800	21	13	16800	2118	1210
1850	23	13	16850	2132	1217
1900	24	14	16900	2145	1225
1950	25	15	16950	2159	1232
2000	27	16	17000	2172	1240
2050	28	16	17050	2186	1247
2100	29	17	17100	2200	1255
2150	31	18	17150	2213	1263
2200	32	19	17200	2227	1270
2250	34	20	17250	2241	1278
2300	35	21	17300	2255	1286
2350	37	22	17350	2269	1293
2400	39	23	17400	2283	1301
2450	40	24	17450	2297	1309
2500	42	25	17500	2311	1317
2550	44	26	17550	2325	1325

A	B	C	A	B	C
2600	45	27	17600	2339	1332
2650	47	28	17650	2353	1340
2700	49	29	17700	2368	1348
2750	51	30	17750	2382	1356
2800	53	31	17800	2396	1364
2850	55	32	17850	2411	1372
2900	57	34	17900	2425	1380
2950	59	35	17950	2440	1388
3000	61	36	18000	2454	1396
3050	63	37	18050	2469	1404
3100	65	38	18100	2483	1412
3150	67	40	18150	2498	1420
3200	69	41	18200	2513	1428
3250	71	42	18250	2527	1437
3300	74	44	18300	2542	1445
3350	76	45	18350	2557	1453
3400	78	46	18400	2573	1461
3450	81	48	18450	2587	1470
3500	83	49	18500	2602	1478
3550	86	51	18550	2617	1486
3600	88	52	18600	2632	1495
3650	91	54	18650	2648	1503
3700	93	55	18700	2663	1511
3750	96	57	18750	2678	1520
3800	98	58	18800	2693	1528
3850	101	60	18850	2709	1537
3900	104	61	18900	2724	1545
3950	106	63	18950	2740	1554
4000	109	65	19000	2755	1562
4050	112	66	19050	2771	1572
4100	115	68	19100	2786	1579
4150	118	70	19150	2802	1588
4200	120	71	19200	2818	1597
4250	123	73	19250	2834	1605
4300	126	75	19300	2849	1614
4350	129	77	19350	2865	1623
4400	132	78	19400	2881	1632
4450	136	80	19450	2897	1640
4500	139	82	19500	2913	1649

A	B	C	A	B	C
4550	142	84	19550	2929	1658
4600	145	86	19600	2946	1667
4650	148	88	19650	2962	1676
4700	152	90	19700	2978	1685
4750	155	91	19750	2994	1694
4800	158	93	19800	3011	1703
4850	162	95	19850	3027	1712
4900	165	97	19900	3043	1721
4950	168	99	19950	3060	1730
5000	172	102	20000	3076	1739
5050	175	104	20050	3093	1748
5100	179	106	20100	3110	1757
5150	183	108	20150	3126	1766
5200	186	110	20200	3143	1775
5250	190	112	20250	3160	1784
5300	194	114	20300	3177	1794
5350	197	116	20350	3194	1803
5400	201	119	20400	3211	1812
5450	205	121	20450	3228	1821
5500	209	123	20500	3245	1831
5550	213	126	20550	3262	1840
5600	217	128	20600	3279	1849
5650	221	130	20650	3296	1859
5700	225	132	20700	3313	1868
5750	229	135	20750	3331	1878
5800	233	137	20800	3348	1887
5850	237	140	20850	3366	1897
5900	241	142	20900	3383	1906
5950	245	145	20950	3401	1916
6000	250	147	21000	3418	1925
6050	254	150	21050	3436	1935
6100	258	152	21100	3453	1944
6150	262	155	21150	3471	1954
6200	267	157	21200	3489	1964
6250	271	160	21250	3507	1974
6300	276	162	21300	3525	1983
6350	280	165	21350	3543	1993
6400	285	168	21400	3561	2003
6450	289	170	21450	3579	2013

A	B	C	A	B	C
6500	294	173	21500	3597	2022
6550	299	176	21550	3615	2032
6600	303	178	21600	3633	2042
6650	308	181	21650	3651	2052
6700	313	184	21700	3670	2062
6750	318	187	21750	3688	2072
6800	322	190	21800	3707	2082
6850	327	192	21850	3725	2092
6900	332	195	21900	3744	2102
6950	337	198	21950	3762	2112
7000	342	201	22000	3781	2122
7050	347	204	22050	3799	2132
7100	352	207	22100	3818	2143
7150	357	210	22150	3837	2153
7200	363	213	22200	3856	2163
7250	368	216	22250	3875	2173
7300	373	219	22300	3894	2183
7350	378	222	22350	3913	2194
7400	384	225	22400	3932	2204
7450	389	228	22450	3951	2214
7500	394	231	22500	3970	2225
7550	400	235	22550	3989	2235
7600	405	238	22600	4009	2246
7650	411	241	22650	4028	2256
7700	416	244	22700	4047	2267
7750	422	247	22750	4067	2277
7800	427	251	22800	4086	2288
7850	433	254	22850	4106	2298
7900	439	257	22900	4125	2309
7950	444	261	22950	4145	2319
8000	450	264	23000	4165	2330
8050	456	267	23050	4185	2341
8100	462	271	23100	4204	2351
8150	468	274	23150	4224	2362
8200	474	278	23200	4244	2373
8250	480	281	23250	4264	2383
8300	486	285	23300	4284	2394
8350	492	288	23350	4304	2405
8400	498	292	23400	4325	2416

A	B	C	A	B	C
8450	504	295	23450	4345	2427
8500	510	299	23500	4365	2438
8550	516	302	23550	4385	2449
8600	523	306	23600	4406	2460
8650	529	310	23650	4426	2471
8700	535	313	23700	4447	2482
8750	542	317	23750	4467	2493
8800	548	321	23800	4488	2504
8850	554	324	23850	4509	2515
8900	561	328	23900	4528	2526
8950	567	332	23950	4550	2537
9000	574	336	24000	4571	2548
9050	581	339	24050	4592	2559
9100	587	343	24100	4613	2571
9150	594	347	24150	4634	2582
9200	601	352	24200	4655	2593
9250	607	355	24250	4676	2604
9300	614	359	24300	4697	2616
9350	621	363	24350	4718	2627
9400	628	367	24400	4740	2639
9450	635	371	24450	4761	2650
9500	642	375	24500	4782	2661
9550	649	379	24550	4804	2673
9600	656	383	24600	4825	2684
9650	663	387	24650	4847	2696
9700	670	391	24700	4869	2707
9750	677	395	24750	4890	2719
9800	685	399	24800	4912	2731
9850	692	404	24850	4934	2742
9900	699	408	24900	4956	2754
9950	706	412	24950	4978	2766
10000	714	416	25000	5000	2777
10050	721	420	25050	5022	2789
10100	729	425	25100	5044	2801
10150	736	429	25150	5066	2813
10200	744	433	25200	5088	2824
10250	751	438	25250	5110	2836
10300	759	442	25300	5133	2848
10350	767	446	25350	5155	2860

A	B	C	A	B	C
10400	774	451	25400	5177	2872
10450	782	455	25450	5200	2884
10500	790	460	25500	5222	2896
10550	798	464	25550	5245	2908
10600	806	469	25600	5268	2920
10650	813	473	25650	5290	2932
10700	821	478	25700	5313	2944
10750	829	483	25750	5336	2956
10800	837	487	25800	5359	2968
10850	846	492	25850	5382	2981
10900	854	496	25900	5405	2993
10950	862	501	25950	5428	3005
11000	870	506	26000	5451	3017
11050	878	510	26050	5474	3030
11100	887	515	26100	5498	3042
11150	895	520	26150	5521	3054
11200	903	525	26200	5544	3067
11250	912	530	26250	5568	3079
11300	920	534	26300	5591	3092
11350	929	539	26350	5615	3104
11400	937	544	26400	5638	3116
11450	946	549	26450	5662	3129
11500	954	554	26500	5686	3142
11550	963	559	26550	5710	3154
11600	972	564	26600	5733	3167
11650	981	569	26650	5757	3179
11700	989	574	26700	5781	3192
11750	998	579	26750	5805	3205
11800	1007	584	26800	5829	3217
11850	1016	589	26850	5854	3230
11900	1025	594	26900	5878	3243
11950	1034	599	26950	5902	3256
12000	1043	605	27000	5926	3269
12050	1052	610	27050	5951	3281
12100	1061	615	27100	5975	3294
12150	1070	620	27150	6000	3307
12200	1080	625	27200	6024	3320
12250	1089	631	27250	6049	3333
12300	1098	636	27300	6074	3346

A	B	C	A	B	C
12350	1108	641	27350	6098	3359
12400	1117	647	27400	6123	3372
12450	1126	652	27450	6148	3385
12500	1136	657	27500	6173	3398
12550	1145	663	27550	6198	3412
12600	1155	668	27600	6223	3425
12650	1165	674	27650	6248	3438
12700	1174	679	27700	6273	3451
12750	1184	685	27750	6299	3464
12800	1194	690	27800	9324	3478
12850	1203	696	27850	6349	3491
12900	1213	701	27900	6375	3504
12950	1223	707	27950	6400	3518
13000	1233	713	28000	6426	3531
13050	1243	718	28050	6451	3544
13100	1253	724	28100	6477	3558
13150	1263	730	28150	6503	3571
13200	1273	735	28200	6529	3585
13250	1283	741	28250	6554	3598
13300	1294	747	28300	6580	3612
13350	1304	753	28350	6606	3626
13400	1314	758	28400	6632	3639
13450	1324	764	28450	6659	3653
13500	1335	770	28500	6685	3667
13550	1345	776	28550	6711	3680
13600	1356	782	28600	6737	3694
13650	1366	788	28650	6764	3708
13700	1377	794	28700	6790	3722
13750	1387	800	28750	6817	3735
13800	1398	806	28800	6843	3749
13850	1408	812	28850	6870	3763
13900	1419	818	28900	6896	3777
13950	1430	824	28950	6923	3792
14000	1441	830	29000	6950	3805
14050	1452	836	29050	6977	3819
14100	1462	842	29100	7004	3833
14150	1473	848	29150	7031	3847
14200	1484	855	29200	7058	3861
14250	1495	861	29250	7085	3875

A	B	C	A	B	C
14300	1506	867	29300	7112	3889
14350	1518	873	29350	7139	3904
14400	1529	880	29400	7167	3918
14450	1540	886	29450	7194	3932
14500	1551	892	29500	7221	3946
14550	1562	899	29550	7249	3961
14600	1574	905	29600	7277	3975
14650	1585	911	29650	7304	3989
14700	1597	918	29700	7332	4004
14750	1608	924	29750	7360	4018
14800	1620	931	29800	7388	4032
14850	1631	937	29850	7415	4047
14900	1643	944	29900	7443	4061
14950	1654	950	29950	7471	4076
15000	1666	957	30000	7500	4090

APPENDIX 3

The Biological Effects of Ionising Radiations

DOSE EFFECTS

Dose (RADS)	*Effects*
10^5	Spastic seizures: death in minutes.
10^4	Damage to central nervous system: death in hours.
10^3	Circulatory and intestinal changes: death in days.
10^2	Radiation sickness, decreased life expectancy and disease resistance, possible sterility.
10	Cataracts and skin conditions, foetal effects and other less obvious effects.

DOSE RATE EFFECTS

Dose Rate (RADS/Day)	*Effects*
10	Debility in 3-6 weeks, death in 3-6 months.
1	Debility in 3-6 months, death in 3-6 years.
10^{-1}	Reduced life expectancy, symptoms appear after several years. Maximum permissible dose in 1930-50.
10^{-2}	Currently acceptable maximum dose rate.
10^{-3}	Natural radiation dose rate.

APPENDIX 4

Photographic Emulsions used in Autoradiography of Biological Materials

Type	Sensitivity	Resolution
No screen X-ray	Good	Poor
Stripping film	Low	Good
Panchromatic film	Medium	Poor
Liquid emulsions	Medium	Good

APPENDIX 5

Characteristics of some Important Isotopes

Isotope Symbol	Element	Most abundant elemental form	Half-life	Principal emissions (MeV)	
^2_1H	Hydrogen	^1_1H (99·9%)	Not radioactive		
^3_1H			12·26 a	β^-	0·0186
$^{14}_6C$	Carbon	$^{12}_6C$ (98·9%)	5570 a	β^-	0·158
$^{15}_7N$	Nitrogen	$^{14}_7N$ (99·6%)	Not radioactive		
$^{18}_8O$	Oxygen	$^{16}_8O$ (99·8%)	Not radioactive		
$^{22}_{11}Na$	Sodium	$^{23}_{11}Na$ (100%)	2·58 a	β^+	0·54 (90%)
					0·51 (10%)
				γ	1·28 (100%)
$^{24}_{11}Na$			15·0 h	β^-	1·39 (100%)
				γ	1·37 (100%)
					2·75 (100%)
$^{32}_{15}P$	Phosphorus	$^{31}_{15}P$ (100%)	14·3 d	β^-	1·71
$^{33}_{15}P$			25 d	β^-	0·25
$^{35}_{16}S$	Sulphur	$^{32}_{16}S$ (95·0%)	87 d	β^-	0·167
$^{40}_{19}K$	Potassium	$^{39}_{19}K$ (93·1%)	$1·27 \times 10^9$ a	β^-	1·32 (89%)
				γ	1·46 (11%)
$^{45}_{20}Ca$	Calcium	$^{40}_{20}Ca$ (97·0%)	163 d	β^-	0·25
$^{51}_{24}Cr$	Chromium	$^{52}_{24}Cr$ (83·7%)	27·8 d	EC	0·76 (100%)
				γ	0·32 (10%)
$^{55}_{26}Fe$	Iron	$^{56}_{26}Fe$ (91·7%)	2·6 a	EC	0·22
$^{59}_{26}Fe$			45 d	β^-	0·46 (53%)
					0·27 (46%)
				γ	1·10 (56%)
					1·29 (44%)

$^{60}_{27}\text{Co}$	Cobalt	$^{59}_{27}\text{Co}$ (100%)	5·27 a	β^-	0·31 (100%)
				γ	1·17 (100%)
					1·33 (100%)
$^{125}_{53}\text{I}$	Iodine	$^{127}_{53}\text{I}$ (100%)	60·0 d	γ	0·035
$^{128}_{53}\text{I}$			25 m	β^-	2.12 (76%)
					etc
				γ	0·45 etc
$^{129}_{53}\text{I}$			$1·6 \times 10^7$ a	β^-	0·15
					0·40
$^{131}_{53}\text{I}$			8·06 d	β^-	0·61 (87%)
					etc
				γ	0·36 (82%)
					0·64 (9%)
$^{133}_{54}\text{Xe}$	Xenon	$^{132}_{54}\text{Xe}$ (27·0%)	5·3 d	β^-	0·35 (99%)
				γ	0·08 (100%)
$^{137}_{55}\text{Cs}$	Caesium	$^{133}_{55}\text{Cs}$ (100%)	30 a	β^-	0·51 (92%)
					1·17 (8%)
$^{198}_{79}\text{Au}$	Gold	$^{197}_{79}\text{Au}$ (100%)	2·7 d	β^-	0·96 (99%)
					0·28 (1%)
				γ	0·41 (99%)

Note:
(i) Some nuclides decay solely by one mechanism, whereas for others various proportions of atoms (indicated as percentages) decay in various ways.
(ii) EC indicates a type of radioactive decay known as electron capture.

APPENDIX 6

The Decay of ^{32}P

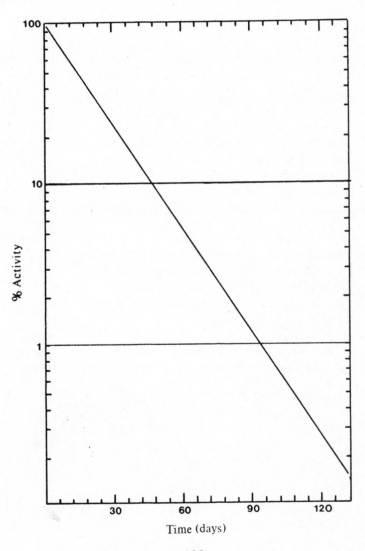

Time (days)

APPENDIX 7

Radioisotope Laboratory Safety Rules

The safe use of radioactive isotopes in teaching experiments has been discussed in some detail in Chapter 6. However a concise list of safety rules for use in radioisotope laboratories is included here for easy reference and possible detachment for use outside this book.

1. No unnecessary materials (books, papers, clothes, etc.) should be taken into the laboratory.

2. Laboratory coats should be worn at all times. The eyes should be protected with goggles and the hands with rubber or disposable gloves. If re-usable gloves are used these should be washed before removal from the hands.

3. Eating, drinking, smoking, and the application of cosmetics is forbidden within the laboratory.

4. All wounds on the wrists and hands should be reported to the class supervisor.

5. If high energy emitting radioisotopes are used, or if radioisotopes are used over long periods of time, a film badge should be worn.

6. Routine analysis of the urine is desirable if millicurie quantities of ^{14}C or ^{3}H are being used.

7. Operations involving high energy β-emitting or γ-emitting radioisotopes should be carried out behind screens.

8. Dispensation of radioisotopes of high specific activity, as for example from stock solutions, should be made by the class supervisor.

9. All work should be carried out on surfaces which are easily cleaned and for added protection, covered with absorbent paper.

10. Work involving radioactive gases or powders should be carried out in a fume cupboard or in special apparatus.

11. Mouth operations of all kinds are forbidden, pipette bulbs or other automatic dispensing devices should be used. Adhesive labels or other materials should not be inserted in the mouth.

12. Any spillage occurring during the experiment should be mopped up with absorbent paper and the area washed. The paper is treated as contaminated waste.

13. If active solutions are spilled on the skin, it should be well washed with soapy water and the accident reported to the class supervisor.

14. Care should be taken when evaporating solutions on to planchets. The planchet should be stood on absorbent paper and a low level of heat intensity used in the final stages in order to avoid spitting of radioactive solutions.

15. Whenever radioactive solutions or materials are carried they should be in a tightly closed container carried within or on another container to minimise the possibility of breakage and/or spillage.

16. To avoid contamination of counting apparatus only correctly prepared samples should be taken into the counting room and rubber gloves removed before handling the counting apparatus.

17. At the conclusion of the experiment all radioactive liquids should be poured into clearly labelled bottles kept for the purpose and never poured down the sink. Radioactive solid waste should be placed in marked pedal bins or plastic bags.

18. Radioactive glassware and apparatus should be well washed and if possible kept separate from non-radioactive materials. This preferably should not be the responsibility of the students.

Planchets and plastic scintillator bottles should be treated as disposable.

19. Radioactive samples that are to be stored for further use should be placed in tightly stoppered containers labelled with yellow isotope warning tape. They should be labelled to indicate the type(s) of isotope involved, the approximate activity, the date of placing in store, and the owner's name.

They should be placed carefully so that they can not be knocked over and broken.

20. On leaving the laboratory the hands should be washed and personal clothing monitored for contamination by a suitable hand monitor.

21. The area over which the experiment was performed should be monitored to check for contamination as a final act before leaving the laboratory.

BIBLIOGRAPHY

GENERAL

BRODA, E. (1960). *Radioactive Isotopes in Biochemistry*. Elsevier.

CHASE, G., and RABINOWITCH, J. (1967) *Principles of Radioisotope Methodology*. Burgess.

GOULDING, K. (1975). Radioisotope Techniques, Ch. 6, in *Principles and Techniques of Practical Biochemistry*. Williams, B. L., Wilson, K. Arnold.

HENDEE, W. R. (1973). *Radioactive Isotopes in Biological Research*. Wiley.

*HORNSEY, D. (1974). Radioactivity and the Biology Teacher. *J. Biol. Educ.* **8**, 313.

HORROCKS, D. (1974). *Applications of Liquid Scintillation Counting*. Academic Press.

INTERNATIONAL ATOMIC ENERGY AGENCY (I.A.E.A.). Technical Report Series. A wide range of advanced studies on the application of radiations and isotopes in biology and other sciences.

KOBAYASHI, T., and MAUDESLEY, D. (1974). *Biological Applications of Liquid Scintillation Counting*. Academic Press.

LAWRENCE, J., MANOWITZ, B., and LOEB, B. (1970). *Radioisotopes and Radiation*. Dover Publications, New York.

NELSON, N., and RUST, J. (1965). The use of Ionising Radiation for Measuring Biological Phenomena, in *Methods of Animal Experimentation*, Vol. II, p. 59. ed. Gay, W.I., Academic Press.

OLIVER, R. (1971). *Principles of the Use of Radioisotope Tracers in Clinical Research Investigations*. Pergamon.

*PAICE, P. (1968). Radioisotopes in School Biology, *School Science Review*, p. 62.

*PARRY WILLIAMS, J., and SERVANT, D. (1968). Radioisotopes in School Biology, *School Science Review*, **49**, 611.

QUAYLE, J. R. (1972). The Use of Isotopes in Tracing Metabolic Paths, p. 157, Vol. VIB. *Methods of Microbiology*, ed. Norris, J. R. and Ribbons, D. W.

ROGINSKI, S., and SCHNOL, S. (1965). *Isotopes in Biochemistry*. Academy of Sciences of USSR.

THE RADIOCHEMICAL CENTRE (1966). *The Radiochemical Manual.* 2 vols. Amersham.

THORNBURN, C. C. (1972). *Isotopes and Radiation in Biology.* Butterworths.

WANG, C., and WILLIS, D. (1965). *Radiotracer Methodology in Biological Science,* Prentice Hall.

WELCH, T., POTCHEN, E., and WELCH, M. (1972). *Fundamentals of the Tracer Method,* Saunders.

WOLF, G. (1964). *Isotopes in Biology,* Academic Press.

RADIATION BIOLOGY

ALTMAN, K. I., GERBER, G. B., and OKADA, S. (1970). *Radiation Biochemistry,* 2 vols. Academic Press.

BACQ, Z. M., and ALEXANDER, P. (1961). *Fundamentals of Radiobiology,* Pergamon.

CASARETT, A. R. (1968). *Radiation Biology,* Prentice Hall.

*COGGLE, J. E. (1971). *Biological Effects of Radiation,* Wykeham Publications.

DERTINGER, H., and JUNG, H. (1969). *Molecular Radiation Biology,* Springer-Verlag.

FABRIKANT, J. (1972). *Radiobiology,* Chicago Year Book Medical Publishers.

*LAWRENCE, C. W. (1971). Cellular Radiobiology, *Studies in Biology,* No. 30. Arnold.

PRASAD, K. (1974). *Human Radiation Biology,* Harper and Row.

SUZUKI, K. (1971). Radiation Biology of Micro-organisms, in *Selected Papers in Biochemistry,* Vol. IV. University Park Press.

WHITSON, G. (1972). *Concepts in Radiation Cell Biology,* Academic Press.

TECHNIQUES (often with some theoretical introduction).

CROSBIE, G. W. (1972). Ionisation Methods of Counting Radioisotopes, p. 65, Vol. VIB. *Methods of Microbiology,* ed. Norris, J., and Ribbons, D. W.

FAIRES, R., and PARKS, B. (1973). *Radioisotope Laboratory Techniques,* 2nd Edition. Butterworths.

GAHAN, P. (1972). *Autoradiography for Biologists,* Academic Press.

GUDE, W. D. (1968). *Autoradiographic Techniques: Localisation of Radioisotopes in Biological Material,* Prentice Hall.

HASH, J. (1972). Liquid Scintillation Counting in Microbiology, p. 109, Vol. 6B. *Methods of Microbiology,* ed. Norris, J. R., and Ribbons, D. W.

INTERTECHNIQUE LTD. PORTSLADE, Sussex. *Digitechniques.* A range of technical reviews mainly on practical aspects of scintillation counting.

KOCH-LIGHT LABORATORIES LTD., Colnbrook, Bucks. Several of the range of booklets produced by this company are relevant to the principles and practice of using radioisotopes.

NUCLEAR CHICAGO. *The Nucleus.*

PACKARD. Technical Bulletins.

ROGERS, A. W. (1967). *Techniques of Autoradiography*, Elsevier.

THE RADIOCHEMICAL CENTRE, Amersham. *Radiochemical Centre Reviews.* A wide range of free booklets concerning applications of isotopes and radiations in biology and medicine.

EXPERIMENTS (often with some theoretical introduction).

*ANDREWS, J., and HORNSEY, D. (1972). *Basic Experiments with Radioisotopes (for Courses in Physics, Chemistry and Biology)*, Pitman Publishing.

*DANCE, J. B. (1973). *Radioisotope Experiments for Schools and Colleges*, 2nd Edition. Pergamon Press.

*PARRY WILLIAMS, J., and SERVANT, D. (1970). Introductory Laboratory Exercises in Radiobiology, *J. Biol. Educ.* **4**, 235.

SAFETY

COMMITTEE OF VICE-CHANCELLORS AND PRINCIPALS OF THE UNIVERSITIES OF THE UNITED KINGDOM. (1966). *Radiological Protection in Universities.*

DEPARTMENT OF EDUCATION AND SCIENCE. (1976). *The Use of Ionising Radiations in Schools, Establishments of Further Education and Teacher Training Colleges.* Administration Memorandum 2/76.

MINISTRY OF LABOUR. (1974). *Code of Practice for the Protection of Persons Exposed to Ionising Radiations in Research and Teaching.* HMSO.

NELSON, N., and RUST, J. (1965). Radiation Hygiene, in *Methods of Animal Experimentation*, Vol. II, p. 1, ed. Gay, W. I. Academic Press.

NOTES

Items marked * are particularly useful for schools.

It is well worth while scanning educational journals in all the sciences, since useful experiments and discussion appear in them from time to time. These journals are: *The Biologist, Journal of Biological Education; Journal of Chemical Education; Chemical Education; Physics Education; School Science Review.*

A number of the more important articles relevant to radiobiology have been included in this bibliography.

INDEX

dilution analysis, 46
ecological studies, 59
enzyme studies, 48
metabolic studies, 51
nuclear activation, 61
physiological studies, 55

radio-immuno assay, 46
sterilisation, 61
turn-over rates, 58

Yeast uptake of ^{32}P, 137